Communications
in Computer and Information Science   2113

Rationale

The CCIS series is devoted to the publication of proceedings of computer science conferences. Its aim is to efficiently disseminate original research results in informatics in printed and electronic form. While the focus is on publication of peer-reviewed full papers presenting mature work, inclusion of reviewed short papers reporting on work in progress is welcome, too. Besides globally relevant meetings with internationally representative program committees guaranteeing a strict peer-reviewing and paper selection process, conferences run by societies or of high regional or national relevance are also considered for publication.

Topics

The topical scope of CCIS spans the entire spectrum of informatics ranging from foundational topics in the theory of computing to information and communications science and technology and a broad variety of interdisciplinary application fields.

Information for Volume Editors and Authors

Publication in CCIS is free of charge. No royalties are paid, however, we offer registered conference participants temporary free access to the online version of the conference proceedings on SpringerLink (http://link.springer.com) by means of an http referrer from the conference website and/or a number of complimentary printed copies, as specified in the official acceptance email of the event.

CCIS proceedings can be published in time for distribution at conferences or as postproceedings, and delivered in the form of printed books and/or electronically as USBs and/or e-content licenses for accessing proceedings at SpringerLink. Furthermore, CCIS proceedings are included in the CCIS electronic book series hosted in the SpringerLink digital library at http://link.springer.com/bookseries/7899. Conferences publishing in CCIS are allowed to use Online Conference Service (OCS) for managing the whole proceedings lifecycle (from submission and reviewing to preparing for publication) free of charge.

Publication process

The language of publication is exclusively English. Authors publishing in CCIS have to sign the Springer CCIS copyright transfer form, however, they are free to use their material published in CCIS for substantially changed, more elaborate subsequent publications elsewhere. For the preparation of the camera-ready papers/files, authors have to strictly adhere to the Springer CCIS Authors' Instructions and are strongly encouraged to use the CCIS LaTeX style files or templates.

Abstracting/Indexing

CCIS is abstracted/indexed in DBLP, Google Scholar, EI-Compendex, Mathematical Reviews, SCImago, Scopus. CCIS volumes are also submitted for the inclusion in ISI Proceedings.

How to start

To start the evaluation of your proposal for inclusion in the CCIS series, please send an e-mail to ccis@springer.com.

Henry Han · Erich Baker
Editors

Next Generation Data Science

Second Southwest Data Science Conference, SDSC 2023
Waco, TX, USA, March 24–25, 2023
Revised Selected Papers

 Springer

Editors
Henry Han (ID)
Baylor University
Waco, TX, USA

Erich Baker
Belmont University
Nashville, TN, USA

ISSN 1865-0929   ISSN 1865-0937 (electronic)
Communications in Computer and Information Science
ISBN 978-3-031-61815-4   ISBN 978-3-031-61816-1 (eBook)
https://doi.org/10.1007/978-3-031-61816-1

This Springer imprint is published by the registered company Springer Nature Switzerland AG
The registered company address is: Gewerbestrasse 11, 6330 Cham, Switzerland

If disposing of this product, please recycle the paper.

Preface

In the era of big data and AI, data science is undergoing an exceptionally rapid evolution, driven by the expansion of big data, advancements in AI, and the emergence of cutting-edge computing technologies such as quantum and neuromorphic computing. Big data serves as the foundational fuel for AI and machine learning models, enabling more precise and robust predictions, deeper insights, and the automation of complex tasks. AI, as a pivotal force, further advances data science by providing more powerful machine learning models that facilitate sophisticated data analysis and enhanced predictive capabilities. Quantum computing notably extends data science's potential with its extraordinary computational power, while neuromorphic computing improves energy efficiency in data science by emulating the human brain's neural architecture. Next-generation data science is emerging through this evolution.

Next-generation data science broadens and deepens existing knowledge and technologies across various domains. Such expansion signifies a shift toward more complex and sophisticated models capable of handling vast and diverse datasets with increased accuracy and efficiency. The broadening and deepening of domains mean that data science principles and techniques are now being applied and customized in new and varied fields through enriched interdisciplinary interaction and collaboration. In the era of next-generation data science, data scientists and AI researchers work closely with domain experts from fields as diverse as medicine, environmental science, fintech, digital humanities, and heliophysics, ensuring that data-driven insights are both accurate and contextually relevant, and are even forging new academic research and industry areas.

Next-generation data science is characterized by its strong ethical nature, driven by the urgent need to address data bias, ensure trustworthiness, and promote ethical AI. Data bias, which leads to distortions and inaccuracies that result in skewed outcomes, often stems from a variety of sources including non-representative sampling, prejudiced measurements, and flawed processing. In particular, mislabeled information can propagate through machine learning models, resulting in inaccurate predictions, unfair practices, and ethical concerns. This issue is especially pronounced in the realm of big data, where the sheer volume and velocity of information can amplify the risks and impacts of bias. The complexity of big data systems often makes it challenging to identify and correct these biases, rendering them more insidious than in smaller, more manageable datasets. Consequently, there is an urgent need for ethical data science techniques that incorporate rigorous ethical protocols to mitigate these issues, ensuring that models are not only technically proficient but also socially responsible and trustworthy. For instance, a college student performance data science analysis would be more credible if factors such as social inequality are incorporated into the design of the machine learning model.

Furthermore, the development of ethical AI necessitates broader applications and more intricate customization across various data science domains to enhance fairness, explainability, and privacy. Fairness ensures that AI models and results avoid various types of biases, including data bias. Explainability seeks to demystify AI models and their decision-making processes, making them clear and understandable to all users and stakeholders. Privacy involves protecting personal information from unauthorized access and preventing privacy leaks in AI methodology design and deployment. For instance, ethical AI techniques designed to promote fairness in high-dimensional data analysis must be specifically adapted for healthcare systems that utilize high-dimensional omics data in the treatment and management of Type II diabetes. Additionally, the explicability of data science systems often depends on the transparency of their machine learning and AI models to prevent 'black-box' learning processes.

On the other hand, while quantum and neuromorphic computing promise significant increases in computing power and efficiency for next-generation data science, they also raise potential ethical concerns. Quantum computing could undermine current encryption methods, potentially making existing data security protocols obsolete and leading to considerable privacy issues if entities with quantum capabilities access sensitive information. Similarly, neuromorphic computing may introduce novel vulnerabilities and security risks. Ethically, it's imperative to safeguard these systems against cyberattacks and ensure they do not harm individuals or society.

Ethical data science will become a cornerstone of next-generation data science domains as the expansion of data-driven decision-making increasingly influences every aspect of society. Incorporating ethical considerations into the design, implementation, and deployment of data science solutions is not merely beneficial but essential. Consequently, next-generation data scientists must not only understand the ethical implications of their work but also actively incorporate these considerations into their algorithms and models. Moreover, ethical data science is likely to encourage greater collaboration across various data science fields, in addition to bringing in legal experts to ensure ethical practices are prioritized. This collaborative effort is crucial for enhancing public trust and driving technological innovation within the realm of next-generation data science.

As we stand on the brink of a new era of next-generation data science, this book invites you to join an exploration of this exciting field.

This book constitutes the proceedings of the Second Southwest Data Science Conference, SDSC 2023, which took place in Waco, Texas, USA, during March 24–25, 2023.

All papers in this book were selected from 72 submissions that underwent a rigorous single-blind review, with each paper being assigned to at least three reviewers. We have accepted 16 full papers and one short paper for this book. The book is divided into two parts: Part 1 focuses on business, social, and foundational data science, while Part 2 covers applied data science, artificial intelligence, and data engineering.

March 2024   Henry Han
                                                                       Erich Baker

Organization

General Chair

Henry Han   Baylor University, USA

Program Committee Chairs

Henry Han	Baylor University, USA
Erich Baker	Belmont University, USA

Steering Committee

Henry Han	Baylor University, USA
Erich Baker	Belmont University, USA
Greg Hamerly	Baylor University, USA
Matthew Fendt	Baylor University, USA
Candace Ditsch	Baylor University, USA
Patrick Hynan	Baylor University, USA

Program Committee

Hisham Al-Mubaid	University of Houston, USA
Mary Lauren Benton	Baylor University, USA
Erdogan Dogdu	Angelo State University, USA
Jeff Forest	Slippery Rock University, USA
Mark Ferguson	University of South Carolina, USA
Michael Gallaugher	Baylor University, USA
Chan Gu	Ball State University, USA
Keith Hubbard	Stephen F. Austin State University, USA
Xiuzhen Hu	Inner Mongolia University of Technology, China
David Kahle	Baylor University, USA
Haiquan Li	University of Arizona, USA
Chun Li	Hainan Normal University, China
David Li	Institute of Plant Physiology and Ecology, CAS, China

Gaoshi Li	Guangxi Normal University, China
Zhen Li	Texas Woman's University, USA
Huiming Liu	Guangxi University, USA
Wenbin Liu	Guangzhou University, China
Michael Pokojovy	University of Texas, El Paso, USA
Eli Olinick	Southern Methodist University, USA
Guimin Qin	Xidian University, China
Greg Speegle	Baylor University, USA
Yang Sun	California Northstate University, USA
James Stamey	Baylor University, USA
Ye Tian	Case Western Reserve University, USA
Stellar Tao	UT Health Center, USA
Tie Wei	Guangxi University, China
Jiacun Wang	Monmouth University, USA
Yi Wu	New York University, USA
Juanying Xie	Shaanxi Normal University, China
Xuemei Yang	Xianyang Normal University, China
Liang Zhao	University of Michigan, Ann Arbor, USA
Jeff Zhang	California State University, Northridge, USA
Joe Zhou	Columbia University, USA

Additional Reviewers

Dongdong Li
Yanbang Zhang
Lihong Yan
Jie Ren
Yutao Fan

Contents

Business, Social and Foundation Data Science

Maximum Likelihood Estimation for Discrete Multivariate Vasicek Processes

Michael Pokojovy[1]([✉]) [iD], Ebenezer Nkum[2] [iD], and Thomas M. Fullerton Jr.[3] [iD]

[1] Department of Mathematics and Statistics, Old Dominion University,
Norfolk, VA 23529, USA
`mpokojovy@odu.edu`
[2] Department of Mathematical Sciences, The University of Texas at El Paso,
El Paso, TX 79968, USA
`enkum@miners.utep.edu`
[3] Department of Economics and Finance, The University of Texas at El Paso,
El Paso, TX 79968, USA
`tomf@utep.edu`

Abstract. Because of low correlation with other asset classes, bonds play major role in portfolio diversification efforts. As investment funds can create robust diversified portfolios with bonds, it is imperative that multiple bonds be analyzed simultaneously. We consider a multivariate extension of the original Vasicek model to multiple zero-coupon bonds. Due to the low-frequency nature of bonds and other debt securities, instead of working in continuous time, we apply the Euler-Maruyama discretization and study the resulting discrete multivariate Vasicek model. We adopt the maximum likelihood estimation (MLE) approach to estimate the parameters of the model, i.e., the long-term mean vector, reversion speed matrix and volatility matrix. Instead of reparametrizing the problem as a VAR(1) model and applying the classical ordinary least squares (OLS) approach for calibration, we rigorously derive a system of nonlinear estimating equations using multivariate vector and matrix calculus and propose a new statistical estimator based on solving this system with a Banach-type fixed point iteration. The performance of our new MLE vs the classical OLS estimator is thoroughly evaluated through a simulation study as well as backtesting analysis performed on 3-month US Treasury and AAA-rated Euro bond yield rates. In both cases, we conclude that our new estimator significantly outperforms the conventional OLS estimator. We also provide a set of Matlab® codes that may prove to be a useful tool for model calibration in connection with portfolio optimization and risk management in the bond market.

Keywords: multivariate Vasicek model · short rates · estimating equations · US Treasuries · Euro bonds

This study received partial funding support from National Science Foundation Grant 2210929, El Paso Water and Raiz Federal Credit Union. Comments and recommendations from SDSC 2023 Chair Professor E. Baker, General Chair Professor H. Han and two anonymous referees are greatly appreciated.

H. Han and E. Baker (Eds.): SDSC 2023, CCIS 2113, pp. 3–18, 2024.
https://doi.org/10.1007/978-3-031-61816-1_1

1   Introduction

Bonds are an important asset class in a well-diversified portfolio due to their low correlation with other asset classes [21]. The study of bond pricing and interest rate dynamics is crucial for portfolio optimization and risk management as it helps investors and analysts better understand the risk and return characteristics of bonds and make informed investment decisions [6]. The well-known Vasicek model, introduced by Oldrich Vasicek in 1977 [30], offers a rational description of the evolution of interest rates over time, specifically for zero-coupon bonds. The model is based on the assumption that interest rates follow a mean-reverting process, i.e., tend to move back towards a long-term average over time. The traditional Vasicek model is a one-factor model, meaning it only considers one source of risk, the interest rate. It is widely used as a benchmark model in the field of mathematical finance and it has been widely applied in the pricing and risk management of bonds and other interest rate derivatives. It is considered a simpler model than other interest rate models, such as the Cox-Ingersoll-Ross (CIR) model [8] and the Hull-White model [18], which are also mean-reverting models but include additional factors. These models are considered to be more general as they can fit to a variety of term structure shapes, however, the Vasicek model is more parsimonious as it requires fewer parameters to be estimated. Another advantage of the Vasicek model is that it has a closed-form solution for the bond prices, which is not the case for many other interest rate models. This makes the Vasicek model more tractable and easier to use in practice.

In recent years, there has been a growing interest in modeling bonds and understanding their dynamics. Some very recent works in this area include, but are not limited to, [5,16,20,22,27], etc. However, there are several limitations associated with most short rate models. One of the main limitations is that traditional bond models, such as the Vasicek model [30], are limited to modeling the behavior of individual bonds. They do not account for simultaneous dynamics of multiple bonds, thus, potentially overlooking dependencies between different bonds. Multivariate extensions of the Vasicek model [11] can effectively be used to account for simultaneous dynamics of multiple bonds and provide a useful tool for portfolio optimization and risk management in the bond market. A number of econometric methods have been developed in recent years to estimate the parameters of multivariate continuous-time models, when the data are only available over a discrete time grid. These methods aim to address the issue of inconsistency caused by ignoring the difference between the continuous-time model and the discretely observed data [23]. Some of the methods include moment-based methods [10,19] as well as simulation-based methods [13,14]. Additionally, there exist nonparametric methods such as density-matching [1], nonparametric regression for approximate moments [28] and Bayesian approaches [12]. Time-varying copulas and reducible stochastic differential equations which account for non-linear features observed in short-term interest rate series were proposed [7] to model multivariate interest rates. However, simulation-based methods require a large number of replications to obtain accurate estimates, which can be computationally intensive. Moment-based methods rely on the assumption that the model is correctly specified, which may not always be the case in practice. Non-parametric methods such as density-matching and nonparametric regression for approximate moments may not be as efficient as parametric methods when the model is correctly specified. Bayesian methods require prior knowledge about the parameters, which may not be known or may be difficult to obtain.

In contrast, maximum likelihood estimation (MLE) is a widely used method that has the advantage of being based on the likelihood function of a parametric model, which is derived from the underlying assumptions. Under appropriate conditions, maximum likelihood (MLE) estimators are efficient, consistent and asymptotically normal. Additionally, MLE is a general method that can be used for a wide range of models and may even work if the model is "slightly" mis-specified. Exact MLE requires computation of the likelihood function, which may be computationally expensive and requires a record of continuous observation of the process. However, a technique for closed-form maximum likelihood estimation (MLE) of multifactor affine yield models with the likelihood function approximated by a series of accurate expansions for the log-likelihood function was developed in [2]. A closed-form approximation to likelihoods for nine affine models, including the Vasicek model, was obtained in [9]. It not clear though if the results can be extended to other affine models. Also, the estimation methods adopted in the research may be sensitive to the choice of initial values for the parameters, which could lead to non-unique or inconsistent estimates. A framework for discretizing linear multivariate continuous time systems was introduced in [31] that leads to a general class of estimators for the mean reversion matrix. They found that the Euler method is useful for estimation of the multivariate models under slow mean reversion. A multivariate extension of the Vasicek model that can model multiple zero-coupon bonds simultaneously was proposed in [11]. The authors, however, used a closed-form maximum likelihood estimation (MLE) technique proposed by [2] to estimate the parameters of the multivariate Vasicek model efficiently, which has its own problems mentioned earlier.

Instead of employing continuous-time multivariate Vasicek processes, whether observed continuously or over a discrete time grid, we propose to analyze the discrete multivariate Vasicek process obtained using the explicit Euler-Maruyama scheme. Our main motivation is the fact that short rate dynamics typically occurs at lower frequencies. Thus, theoretical complications arising from accounting for continuous time is not justified empirically. Focusing on a finite time horizon, whether short or long, we compute the log-likelihood function and explicitly derive a set of estimating equations. Instead of reparameterizing the problem as a VAR(1) (vector autoregression model with lag 1) model and fitting it using the conventional OLS approach, we work with the original set of parameters. We develop a Banach-type fixed point iteration procedure to solve the associated estimating equations. Since our estimator does not involve reparameterization, empirical evidence suggests it performs better at estimating the long-term mean, especially for ill-conditioned or singular reversion speed matrices.

The rest of the paper is structured as follows. In Sect. 2, we introduce the Vasicek model and its multivariate extension as well as finite difference discretization. In Sect. 3, we rigorously derive the system of estimating equations and present our new MLE estimator based on these findings. In Sect. 4.1, we conduct a simulation study to compare the performance between our MLE estimator and the classical OLS estimator. In Sect. 4.2, we illustrate how these estimators can be applied to a bivariate real-world dataset and perform a benchmarking analysis to compare both approaches and highlight their practical relevance and applicability. In Sect. 5, we summarize our study and offer conclusions, drawing together the key insights and contributions of our research. Auxiliary results and proofs are given in Appendix.

2   Vasicek Short Rate Model

The classical Vasicek short rate process $(r_t)_{t \geq 0}$ is given as a solution to stochastic differential equation (SDE):

$$dr_t = \alpha(r^* - r_t)\,dt + \sigma dW_t \text{ for } t \in [0, T], \quad r_0 = r^0, \tag{1}$$

where r^* is the long-term mean, α is the inversion speed, $\sigma > 0$ is the instantaneous or spot volatility and r^0 is the initial rate (typically, nonrandom) [30]. When multiple, say, p short rates $\boldsymbol{R}_t = (R_{t,1}, \ldots, R_{t,p})'$ are to be considered simultaneously, a straightforward multivariate generalization of Eq. (1) is given via the coupled system of multivariate SDEs

$$d\boldsymbol{R}_t = \boldsymbol{A}(\boldsymbol{R}^* - \boldsymbol{R}_t)\,dt + \boldsymbol{\Sigma}^{1/2}\,d\boldsymbol{W}_t \text{ for } t \in [0, T], \quad \boldsymbol{R}_0 = \boldsymbol{R}^0, \tag{2}$$

where $\boldsymbol{R}^* \in \mathbb{R}^p$ is the long-term mean vector, $\boldsymbol{A} \in \mathbb{R}^{p \times p}$ is the mean inversion speed matrix, $\boldsymbol{\Sigma} \in \mathbb{R}^{p \times p}$ is the (positive definite) spot volatility matrix and $(\boldsymbol{W}_t)_{t \geq 0}$ is p-variate standard Wiener process. Here and in the sequel, the matrix root is defined in terms of the spectral theorem.

Since short rates are typically observed at lower frequencies, say, daily, the high-frequency context given by SDEs can adequetately be replaced with a time series. Letting $\Delta t = \frac{T}{n}$ for $n \in \mathbb{N}$ and applying the explicit Euler-Maruyama discretization scheme, we obtain a time series approximation of SDE (2):

$$\tfrac{1}{\Delta t}(\boldsymbol{R}_{t_{j+1}} - \boldsymbol{R}_{t_j}) = \boldsymbol{A}(\boldsymbol{R}^* - \boldsymbol{R}_{t_j}) + \tfrac{1}{\Delta t}\boldsymbol{\Sigma}^{1/2}(\boldsymbol{W}_{t_{j+1}} - \boldsymbol{W}_{t_j}), j = 0, \ldots, n-1 \tag{3}$$

or, equivalently,

$$\boldsymbol{R}_{t_{j+1}} = (\Delta t)\boldsymbol{A}\boldsymbol{R}^* + (\boldsymbol{I}_{p \times p} - (\Delta t)\boldsymbol{A})\boldsymbol{R}_{t_j} + \varepsilon_{t_j}, j = 0, \ldots, n-1 \tag{4}$$

with $\varepsilon_{t_j} = \boldsymbol{\Sigma}^{1/2}(\boldsymbol{W}_{t_{j+1}} - \boldsymbol{W}_{t_j}) \overset{\text{i.i.d.}}{\sim} \mathcal{N}(\boldsymbol{0}_p, (\Delta t)\boldsymbol{\Sigma})$. Given parameters $\boldsymbol{R}^*, \boldsymbol{A}, \boldsymbol{\Sigma}$ (or estimates thereof), Eq. (4) can be used to forecast future short rates.

3   Maximum Likelihood Estimation

A common way to estimate the parameters of Eq. (3) is the maximum likelihood estimation (MLE) method. Instead of directly applying the MLE procedure to

Eq. (3), the classical approach consists in reparametrizing the equation as a p-variate VAR(1) (vector autoregressive) model with lag 1 (cf. Eq. (4))

$$\boldsymbol{R}_{t_{j+1}} = \boldsymbol{c} + \boldsymbol{M}\boldsymbol{R}_{t_j} + \boldsymbol{e}_{t_j} \tag{5}$$

where $\boldsymbol{c} = (\Delta t)\boldsymbol{A}\boldsymbol{R}^*, \quad \boldsymbol{M} = \boldsymbol{I}_{p \times p} - (\Delta t)\boldsymbol{A}$ and $\mathrm{Cov}[\boldsymbol{e}_{t_j}] \equiv \boldsymbol{\Omega} = (\Delta t)\boldsymbol{\Sigma}.$ (6)

In turn, provided $\boldsymbol{A}$ is invertible, the MLE procedure for Eq. (5) is equivalent with applying the OLS approach to regress $\boldsymbol{R}_{t_{j+1}}$ on $\boldsymbol{R}_{t_j}$. Letting

$$\hat{\boldsymbol{c}}_{\mathrm{OLS}}, \ \hat{\boldsymbol{M}}_{\mathrm{OLS}} \ \text{and} \ \hat{\boldsymbol{\Omega}}_{\mathrm{OLS}} \tag{7}$$

denote the OLS estimates (where $\hat{\boldsymbol{\Omega}}_{\mathrm{OLS}}$ does not include the Bessel correction), the original parameters can equivalently be reconstructed via

$$\hat{\boldsymbol{A}}_{\mathrm{OLS}} = \tfrac{1}{\Delta t}(\boldsymbol{I}_{p \times p} - \hat{\boldsymbol{M}}_{\mathrm{OLS}}), \quad \hat{\boldsymbol{R}}^*_{\mathrm{OLS}} = \tfrac{1}{\Delta t}\hat{\boldsymbol{A}}^{-1}_{\mathrm{OLS}}\hat{\boldsymbol{c}}_{\mathrm{OLS}} \ \text{and} \ \hat{\boldsymbol{\Sigma}}_{\mathrm{OLS}} = \tfrac{1}{\Delta}\hat{\boldsymbol{\Omega}}_{\mathrm{OLS}}. \tag{8}$$

If $\boldsymbol{A}$ and/or $\hat{\boldsymbol{A}}$ are singular or ill-conditioned, the OLS estimates from Eq. (7) (especially, $\hat{\boldsymbol{R}}^*_{\mathrm{OLS}}$) may perform very poorly (cf. Sect. 4.1), which may adversely affect our ability to draw practical conclusions from the model since many analyses heavily rely on solid knowledge about the long-term levels.

Instead of following the OLS estimators arising from reparameterization, we propose a direct MLE approach in what follows. Observing

$$(\boldsymbol{R}_{t_{j+1}} - \boldsymbol{R}_{t_j}) - \boldsymbol{A}(\boldsymbol{R}^* - \boldsymbol{R}_{t_j})(\Delta t) \overset{\text{i.i.d.}}{\sim} \mathcal{N}(\boldsymbol{0}_p, (\Delta t)\boldsymbol{\Sigma}), \tag{9}$$

the log-likelihood function $\ell(\boldsymbol{\theta}|\boldsymbol{R})$ (scaled by Δt) associated with the process $\boldsymbol{R} = (\boldsymbol{R}_0, \boldsymbol{R}_1, \ldots, \boldsymbol{R}_n)'$ and parameterized by $\boldsymbol{\theta} := (\boldsymbol{R}^*, \boldsymbol{A}, \boldsymbol{\Sigma})'$ reads as

$$\ell(\boldsymbol{\theta}|\boldsymbol{R}) = (\Delta t) \sum_{j=0}^{n-1} \log \varphi\Big(\Delta \boldsymbol{R}_{t_j} - \boldsymbol{A}(\boldsymbol{R}^* - \boldsymbol{R}_{t_j})\Delta t \, \big| \, \boldsymbol{0}_p, (\Delta t)\boldsymbol{\Sigma}\Big) \tag{10}$$

$$= -\frac{pT}{2} \log\big(2\pi(\Delta t)\big) - \frac{T}{2} \log|\boldsymbol{\Sigma}| \tag{11}$$

$$- \frac{1}{2} \sum_{j=0}^{n-1} \Big(\Delta \boldsymbol{R}_{t_j} - \boldsymbol{A}(\boldsymbol{R}^* - \boldsymbol{R}_{t_j})(\Delta t)\Big)' \boldsymbol{\Sigma}^{-1} \Big(\Delta \boldsymbol{R}_{t_j} - \boldsymbol{A}(\boldsymbol{R}^* - \boldsymbol{R}_{t_j})(\Delta t)\Big), \tag{12}$$

where $\Delta \boldsymbol{R}_{t_j} := \boldsymbol{R}_{t_{j+1}} - \boldsymbol{R}_{t_j}$ and $\varphi(\boldsymbol{x}|\boldsymbol{\mu}, \boldsymbol{\Sigma})$ is the p-variate Gaussian density.

The maximum likelihood estimator is given as

$$\boldsymbol{\theta}_{\mathrm{MLE}} \equiv (\boldsymbol{R}^*_{\mathrm{MLE}}, \boldsymbol{A}_{\mathrm{MLE}}, \boldsymbol{\Sigma}_{\mathrm{MLE}}) = \arg\max_{\boldsymbol{\theta} \in \Theta} \ell(\boldsymbol{\theta}|\boldsymbol{R}) \equiv \arg\max_{(\boldsymbol{R}^*, \boldsymbol{A}, \boldsymbol{\Sigma}) \in \Theta} \ell(\boldsymbol{R}^*, \boldsymbol{A}, \boldsymbol{\Sigma}|\boldsymbol{R})$$

with the (open) parameter set $\Theta = \big\{(\boldsymbol{R}^*, \boldsymbol{A}, \boldsymbol{\Sigma}) \,|\, \boldsymbol{R}^* \in \mathbb{R}^p, \boldsymbol{A} \text{ invertible}, \boldsymbol{\Sigma} \in \mathbb{R}^{p \times p}, \boldsymbol{\Sigma}' = \boldsymbol{\Sigma}, \boldsymbol{\Sigma} \succ 0\big\}$. On the strength of Fermat's interior extremum theorem, local maximization of $\ell(\boldsymbol{\theta}|\boldsymbol{R})$ entails solving the set of estimating equations

$$\boldsymbol{U}(\boldsymbol{\theta}|\boldsymbol{R}) = \boldsymbol{0} \ \text{with the score function} \ \boldsymbol{U}(\boldsymbol{\theta}|\boldsymbol{R}) \equiv \frac{\partial}{\partial \boldsymbol{\theta}} \ell(\boldsymbol{\theta}|\boldsymbol{R}). \tag{13}$$

Theorem 1. *The maximum likelihood estimating Eqs. (13) can equivalently be expressed in the fixed point form:*

$$R^* = \frac{1}{n}\sum_{j=0}^{n-1} R_{t_j} + \frac{1}{T}A^{-1}(R_T - R^0), \tag{14}$$

$$A = \frac{1}{\Delta t}\left(\sum_{j=0}^{n-1}(R_{t_{j+1}} - R_{t_j})(R^* - R_{t_j})'\right)\left(\sum_{j=0}^{n-1}(R^* - R_{t_j})(R^* - R_{t_j})'\right)^{-1}, \tag{15}$$

$$\Sigma = \frac{1}{T}\sum_{j=0}^{n-1}\left(\Delta R_{t_j} - A(R^* - R_{t_j})(\Delta t)\right)\left(\Delta R_{t_j} - A(R^* - R_{t_j})(\Delta t)\right)'. \tag{16}$$

Equations (14)–(16) given by Theorem 1 comprise a system of coupled non-linear vector/matrix equations. The applicability of off-the-shelf solvers for Eqs. (14)–(16) is limited in our context as neither of them employ an adequate error measure for matrix arguments (unlike our Eq. (17)). Instead, we propose a simple and candid computational algorithm based on Banach's fixed point theorem.

Algorithm 1: Fixed-Point MLE

Data: Observed short rate vectors $R_{t_0}, \ldots, R_{t_n}$, time step $\Delta t = \frac{T}{n}$, tolerance level $\varepsilon > 0$, maximal number of iterations `MaxIter`

1 Initialize $R^* \leftarrow \frac{1}{n}\sum_{j=0}^{n-1} R_{t_j}$

2 Plug R^* into Equations (15) and (16) to compute A and Σ

3 **for** $k = 0, 1, \ldots, \texttt{MaxIter}$ **do**

4 $\quad$ Compute:

5 $\quad \tilde{R}^* \leftarrow \frac{1}{n}\sum_{j=0}^{n-1} R_{t_j} + \frac{1}{T}A^{-1}(R_T - R^0),$

6 $\quad \tilde{A} \leftarrow \frac{1}{\Delta t}\left(\sum_{j=0}^{n-1}(R_{t_{j+1}} - R_{t_j})(R^* - R_{t_j})'\right)\left(\sum_{j=0}^{n-1}(R^* - R_{t_j})(R^* - R_{t_j})'\right)^{-1},$

7 $\quad \tilde{\Sigma} \leftarrow \frac{1}{T}\sum_{j=0}^{n-1}\left(\Delta R_{t_j} - A(R^* - R_{t_j})(\Delta t)\right)\left(\Delta R_{t_j} - A(R^* - R_{t_j})(\Delta t)\right)'$

8 $\quad$ Compute the squared error

9 $\quad \epsilon^2 \leftarrow \frac{1}{p}\|R^*_{(k)} - R^*_{(k-1)}\|^2 + \frac{1}{p^2}\rho^2(A_{(k-1)}, A_{(k)}) + \frac{1}{p^2}\rho^2(\Sigma_{(k-1)}, \Sigma_{(k)})$

$$\text{with } \rho^2(U, V) = \|I_{p \times p} - U^{-1}V\|_{\mathcal{F}}^2 + \|I_{p \times p} - UV^{-1}\|_{\mathcal{F}}^2. \tag{17}$$

10 $\quad$ Update $R^* \leftarrow \tilde{R}^*$, $A \leftarrow \tilde{A}$ and $\Sigma \leftarrow \tilde{\Sigma}$

11 $\quad$ **if** $\epsilon^2 < \varepsilon$ **then**

12 $\quad\quad$ | break

13 $\quad$ **end**

14 **end**

15 **return** R^*, A, Σ

For numerical stability, the usual matrix inverse $(\cdot)^{-1}$ in Algorithm 1 is replaced with Moore-Penrose pseudoinverse in our algorithmic implementation.

Here, $\| \cdot \|_{\mathcal{F}}$ denotes the norm induced by the usual Frobenius scalar product (cf. Supplemental Lemma S2.4). Since A does not need to be symmetric, instead of the more conventional Riemannian metric used to quantify the distance between symmetric matrices, we employed the expression in Eq. (17) to account for potential lack of symmetry and/or positive definiteness. A Bessel correction factor $\frac{n}{n-p-1}$ can be introduced in Eq. (16) for unbiasedness [15].

The crucial observation for the "warmstart" in Algorithm 1, step 1 and 2, is the following consideration. Assuming $\frac{1}{T}A^{-1}(R_T - R^0)$ as $T \to \infty$ in appropriate probabilistic sense, which at minimum necessitates A to be positive definite, the latter term on the right-hand side of Eq. (14) vanishes, yielding $R^* \approx \frac{1}{n}\sum_{j=0}^{n-1} R_{t_j}$ and causing the equations to decouple in the limit.

Remark 1. Assuming the arithmetic with floating point numbers has an $\mathcal{O}(1)$ complexity, the initialization step has the complexity of $\mathcal{O}(np)$. At each of up to `MaxIter` iterations of Algorithm 1, we have:

- The complexity of computing $\tilde{R}$ is $\mathcal{O}(np + p^3)$. This is due to the fact the Moore-Penrose pseudoinverse used in the algorithm is obtained via singular value decomposition, which, in turn, has the complexity of $\mathcal{O}(p^3)$. This dominates the complexity of $\mathcal{O}(p^2)$ of multiplying a p-dimensional vector by a $p \times p$ matrix.
- The complexity of computing $\tilde{A}$ is $\mathcal{O}(np^2 + p^3)$ since Moore-Penrose pseudoinverse computation is involved. The latter dominates the complexity of any traditional textbook matrix multiplication algorithm.
- The complexity of computing $\tilde{\Sigma}$ is only $\mathcal{O}(np^2)$ since no pseudo-inverse computation or matrix by matrix multiplication is involved.
- Finally, calculating ϵ^2 requires $\mathcal{O}(p^3)$ operations due to the necessity of computing pseudo-inverses. This dominates the complexity of both matrix-by-matrix multiplication and Frobenius norm computation $(\mathcal{O}(p^2))$.

In sum, we end up with an overall complexity of $\mathcal{O}(\texttt{MaxIter} \cdot (np^2 + p^3))$. Since p is typically very moderate, the cubic complexity in p does not impose major computational restrictions. Practically, selecting the tolerance $\varepsilon = 10^{-8}$, the algorithm usually terminated after a few dozen iterations in our simulation study (viz. Sects. 4.1 and 4.2).

As it is generally the case for MLE estimation, our MLE estimator has a variety of attractive properties such as asymptotic efficiency, asymptotic consistency and asymptotic normality. While typically resistant to minor model violations or presence of few scattered outliers, MLE estimators are not robust and tend to breakdown under less favorable circumstances [17, 26]. This shortcoming is also shared by OLS estimators.

Although beyond the scope of the present paper, we envision several possible ways of rendering Algorithm 1 robust. One way to robustify the MLE procedure is to replace it with minimizing the density power divergence functional [4], which is expected to lead to a system of estimating equations similar to Eq. (14)–(16) but containing nonlinear weights diminishing on the tails. Another approach is to maximize a trimmed log-likelihood with a procedure reminiscent of our recent contribution [24], which would lead to introducing an extra Frank-Wolfe iteration at each step of Algorithm 1. In turn, OLS estimation can potentially be robustified using the well-known MCD regression [25]. We intend to explore these and other aspects in our future work.

4 Simulations and Example

4.1 Simulation Study

We conducted a simulation study to compare the performance of our direct MLE estimator (referred to as "our MLE" in the following) and the classical OLS estimator at estimating the parameters of discrete Vasicek process (3). We considered a hypothetical short rate situation summarized in Table 1. The setup corresponds to a high-interest environment (US/Euro long-term levels of 5% and 6%, respectively) designed to complement the low-interest setup in Sect. 4.2 below. The initial value in Eq. (2) was selected $R^0 = (4, 7)'$.

Table 1. "True" parameter values in a hypothetical high-interest environment.

$$R^* = \begin{pmatrix} 5 \\ 6 \end{pmatrix} \quad A = \frac{1}{365} \begin{pmatrix} 9 & 4 \\ 2 & 3 \end{pmatrix} \quad \Sigma = \frac{1}{365^2} \begin{pmatrix} 7 & 3 \\ 3 & 7 \end{pmatrix}$$

Selecting $\Delta t = 1$ day and a grid of time horizons $T = 60, 90, 120, \ldots, 330, 360$ days, 5,000 daily paths $\{R_{t_j, i}\}_{j=0,\ldots,n}$ for $i = 1, \ldots, 5,000$ of the bivariate discrete Vasicek model in Eq. (2) were generated. For each of the paths, the parameters R^*, A, Σ were estimated using both our MLE and the classical OLS estimators yielding the estimator triples $(\hat{R}^*_{\mathrm{MLE},i}, \hat{A}_{\mathrm{MLE},i}, \hat{\Sigma}_{\mathrm{MLE},i})$ and $(\hat{R}^*_{\mathrm{OLS},i}, \hat{A}_{\mathrm{OLS},i}, \hat{\Sigma}_{\mathrm{OLS},i})$, respectively. The empirical MSEs (mean squared errors) were estimated by:

$$\widehat{\mathrm{MSE}}(\hat{R}^*) = \frac{1}{5,000} \sum_{i=1}^{5,000} \|\hat{R}^*_i - R^*\|^2, \tag{18}$$

$$\widehat{\mathrm{MSE}}(\hat{A}) = \frac{1}{5,000} \sum_{i=1}^{5,000} \rho^2(\hat{A}_i, A), \quad \widehat{\mathrm{MSE}}(\hat{\Sigma}) = \frac{1}{5,000} \sum_{i=1}^{5,000} \rho^2(\hat{\Sigma}_i, \Sigma) \tag{19}$$

represented as a function of the training time horizon T for each of the two estimators. See Eq. (17) for the definition of $\rho(\cdot, \cdot)$. Note that our MSE definition is stronger than the one typically employed as we aim to more adequately measure the quality of matrix parameter estimation.

Figure 1 plots empirical root-MSE for each of the three parameters R^*, A, Σ obtained with our MLE and classical OLS estimators vs training time horizon

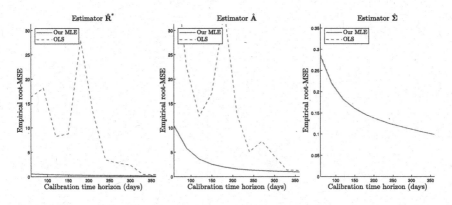

Fig. 1. Simulated root-MSE using MLE and OLS estimators (Color figure online)

size T. All errors generally appear to decrease with increasing T as the Central Limit Theorem for MLE and Gauss-Markov Theorem for OLS estimation predict. For each of the three parameters, our MLE estimator significantly outperformed the OLS one. The advantage was particularly pronounced for estimating R^* and A (left and central panel of Fig. 1), where the empirical root-MSE of our MLE (blue) was much smaller than that of OLS (red). This is due to the fact that the OLS estimator is more likely to produce singular or ill-conditioned estimates that will cause $\widehat{MSE}(\hat{A})$ to explode. No major difference between MLE and OLS was observed for Σ (cf. right panel of Fig. 1). Nonetheless, analyzing the empirical root-MSE ratio (MLE to OLS) in Supplemental Figure S(1), our MLE is seen to outperform the OLS with the advantage dropping from about 1% for smaller calibration horizons T to zero as $T \to \infty$. In sum, our MLE estimator appears to be very beneficial, especially for shorter training horizons, which are typical in turbulent interest rate environments, including the currently prevalent one.

4.2   Example

Two financial datasets were concurrently analyzed in this paper. The first dataset are the Daily Treasury Par Yield Curve Rates provided by the US Department of the Treasury and publicly available online [29]. It contains daily estimates of the yield curve for US Treasury securities with maturities ranging from 1 month to 30 years which is available for each business day. The dataset provides information on the daily yields of US Treasury securities, which are considered to be risk-free investments. The yield curve is a graphical representation of the relationship between the yield on a security and it maturity.

The second dataset is the Daily Euro Par Yield Curve Rates published by the European Central Bank (ECB) [3]. It provides daily yield curves for different maturities of euro-denominated government bonds. The dataset includes data on yields for maturities ranging from 1 month to 30 years and is based on the prices

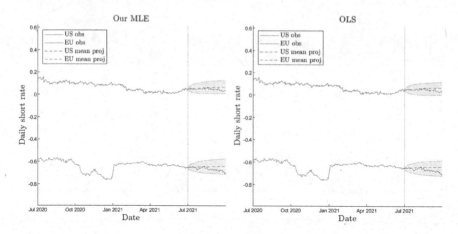

Fig. 2. Historic US/EU 3-month rates (7/1/2020–6/30/2021) as well as forecasted mean and 90% projection bands (7/1/2021–9/30/2021)

of bonds traded in the secondary market. The Euro area yield curve shows separately AAA-rated euro area central government bonds and all Euro area central government bonds (including AAA-rated). It is updated every target business day at noon (12:00 CET). The Euro area yield curves are reported for zero-coupon and forward and par yield curves. The European Central Bank quantifies zero-coupon yield curves for the Euro area and derives forward and par yield curves. The methodology used to estimate the yield rates are provided at the European Central Bank website [3]. These datasets are a valuable tool used by investors, policymakers, and economists to understand the state of the US economy and the Eurozone economy, respectively.

Table 2. Parameter estimates using MLE and OLS estimators.

	$\hat{R}^*$	$\hat{A}$	$\hat{\Sigma}$
Our MLE	$\begin{pmatrix} 0.0629 \\ -0.6534 \end{pmatrix}$	$\begin{pmatrix} 0.0221 \ 0.0035 \\ 0.0034 \ 0.0108 \end{pmatrix}$	$10^{-4} \cdot \begin{pmatrix} 0.5269 \ 0.0253 \\ 0.0253 \ 0.4278 \end{pmatrix}$
OLS	$\begin{pmatrix} 0.0712 \\ -0.6377 \end{pmatrix}$	$\begin{pmatrix} 0.0222 \ 0.0035 \\ 0.0034 \ 0.0108 \end{pmatrix}$	$10^{-4} \cdot \begin{pmatrix} 0.5275 \ 0.0257 \\ 0.0257 \ 0.4282 \end{pmatrix}$

Since we are particularly interested in short rate modeling, we selected 3-month maturities from both the Daily Treasury Par Yield Curve Rates and the AAA-rated euro area from the Daily Euro Par Yield Curve Rates as a bivariate time series to illustrate MLE estimation and argue in favor of our new MLE estimator. To calibrate the multivariate discrete Vasicek model, we selected the time series data for the 7/1/2020–6/30/2021 time period. With both FED rates and Euro area interest rates being next to zero, this corresponds to a rather extreme low-interest environment. The reason for this choice is that there were no major changes in either FED Funds Effective Rate or Main Refinancing Rate for the ECB over this time period rendering the model applicable without the necessity to account for possible change points. The historic data are displayed in Fig. 2 (left and right panel) to the left of the dotted vertical line (7/1/2021). Linearly inter-/extrapolating to account for weekends and holidays, both our MLE and usual OLS estimators were calibrated using 365 days' worth of data.

As part of post hoc diagnostics, using the point estimates presented in Table 2, we computed the sphered (decorrelated standardized) empirical residuals

$$\hat{\zeta}_{t_j} = \hat{\Sigma}^{-1/2}\big((R_{t_{j+1}} - R_{t_j}) - \hat{A}(\hat{R}^* - R_{t_j})(\Delta t)\big) \tag{20}$$

displayed in Fig. 3. Given good agreement between the OLS and MLE estimates in Table 2, respective sphered residuals in Fig. 3 also appear to be nearly identical. Visually inspecting the scatterplots in Fig. 3, vetical "statification" effects can be observed, probably, due to rounding effects in US short rate data. Performing a Šidák's correction for multiple testing $\alpha_{\text{corr}} = 1 - (1-\alpha)^{1/n}$ to attain a (nominal) family-wise error rate of $\alpha = 0.05$, 95%-prediction circles with radius of $\big(\chi^2_{2,1-\alpha_{\text{corr}}}\big)^{1/2} \approx 4.2113$ were used to test for outliers. This leads to only two outlying empirical sphered residuals being detected for both estimators. Despite formal statistical significance, these two outliers do not seem practically relevant since only 2 out of 264 residuals (0.55%) fall outside of prediction limits. Thus, no strong evidence of model violation exists.

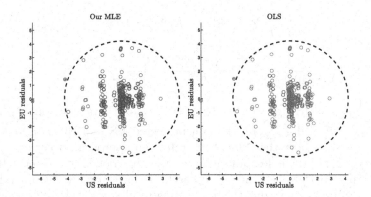

Fig. 3. Sphered empirical residuals for MLE and OLS estimators with respective 95%-prediction circles

For backtesting and benchmarking purposes, 10,000 future projections were simulated according to Eq. (2) using our MLE and usual OLS estimators. The forecasted mean as well as 90%-forecast bands for 7/1/2021–9/30/2021 are displayed to the right of the dotted vertical line on the left and the right panel of

Fig. 2 for MLE and OLS estimates, respectively. No major difference between the two set of predictions can be visually established. In both situations, the observed rates stayed within the 90%-bands for the whole 3-month duration of the forecast. Although, the EU rates started to signal a potential downturn at certain point within the backtesting period.

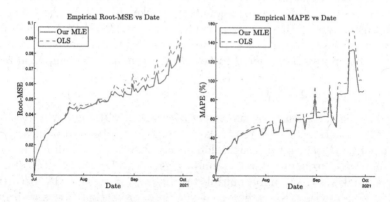

Fig. 4. Root-MSE and MAPE using MLE and OLS estimators

For a more objective comparison between the two methods, we performed a Monte Carlo simulation of size 10,000 to produce empirical root-MSE (mean squared error) and MAPE (mean average percentage error) calculated via

$$\widehat{\mathrm{MSE}}(t_j) = \frac{1}{10{,}000} \sum_{i=1}^{10{,}000} \left(|\hat{R}^{\mathrm{US}}_{t_j,i} - R^{\mathrm{US}}_{t_j,\mathrm{obv}}|^2 + |\hat{R}^{\mathrm{EU}}_{t_j,i} - R^{\mathrm{EU}}_{t_j,\mathrm{obv}}|^2 \right), \qquad (21)$$

$$\widehat{\mathrm{MAPE}}(t_j) = \frac{1}{10{,}000} \sum_{i=1}^{10{,}000} \left(\frac{|\hat{R}^{\mathrm{US}}_{t_j,i} - R^{\mathrm{US}}_{t_j,\mathrm{obv}}|}{|R^{\mathrm{US}}_{t_j,\mathrm{obv}}|} + \frac{|\hat{R}^{\mathrm{EU}}_{t_j,i} - R^{\mathrm{EU}}_{t_j,\mathrm{obv}}|}{|R^{\mathrm{EU}}_{t_j,\mathrm{obv}}|} \right) \qquad (22)$$

where $R^{\mathrm{US}}_{t_j,\mathrm{obv}}$ and $R^{\mathrm{EU}}_{t_j,\mathrm{obv}}$ are the observed US and EU rates at time t_j, respectively, and $\hat{R}^{\mathrm{US}}_{t_j,i}$ and $\hat{R}^{\mathrm{EU}}_{t_j,i}$ are the i-th projected US and EU rate at time t_j, respectively. Since both the US and EU rates can approach zero or even change their sign, the MAPE is less informative than the root-MSE. Nevertheless, we chose to report both as it is the usual practice in measuring and comparing time series performance. Analyzing the backtesting results in Fig. 4, we conclude that forecasting based on our MLE estimator outperforms that using the OLS estimator. The difference would have likely been more pronounced if a shorter time horizon (say, 120 days instead of 365) would have been used.

5   Summary and Conclusions

The discrete multivariate Vasicek model can be used for modeling multiple low-frequency debt securities and other assets simultaneously, which takes into account their simultaneous dynamics and accounts for potential dependencies between and associated credit risk. We used the MLE approach to derive estimating equations and put forth a new estimator for the model parameters, specifically, the long-term mean vector, reversion speed matrix, and volatility matrix. Both our MLE and the usual OLS methods were evaluated and compared through a simulation study based on synthetic data and through backtesting when applied to forecasting 3-month yield rates of US Treasury and AAA-rated Euro bonds. The empirical evidence suggest that our MLE estimator computed by numerically solving the nonlinear system of estimating equations results in improved performance compared to the traditional OLS approach. The study provides a useful algorithmic tool for MLE estimation that can be used in portfolio optimization and risk management in the bond market, which, in turn, can help investors and analysts better understand the risk and return characteristics of bonds and make informed investment decisions. Our future work will include developing a robust version of Algorithm 1. A set of Matlab® codes to reproduce the results from Sects. 4.1 and 4.2 is available at https://github.com/mpokojovy/mvvasicek/. Supplemental figure and auxiliary proofs referred to in the paper are provided in https://github.com/mpokojovy/mvvasicek/blob/main/PDF/supplement.pdf.

A   Proof of Theorem 1

Differentiating with respect to $\boldsymbol{R}^*$ and invoking Supplemental Lemma S2.1

$$\frac{\partial}{\partial \boldsymbol{R}^*}\ell(\boldsymbol{\theta}|\boldsymbol{R}) = (\Delta t)\frac{\partial}{\partial \boldsymbol{R}^*}\sum_{j=0}^{n-1}\left(\boldsymbol{A}'\boldsymbol{\Sigma}^{-1}(\Delta \boldsymbol{R}_{t_j})\right)'(\boldsymbol{R}^* - \boldsymbol{R}_{t_j}) \qquad (23)$$

$$-\frac{(\Delta t)^2}{2}\frac{\partial}{\partial \boldsymbol{R}^*}\sum_{j=0}^{n-1}\left((\boldsymbol{A}'\boldsymbol{\Sigma}^{-1}\boldsymbol{A})(\boldsymbol{R}^* - \boldsymbol{R}_{t_j})\right)'(\boldsymbol{R}^* - \boldsymbol{R}_{t_j}) \qquad (24)$$

$$= (\Delta t)(\boldsymbol{\Sigma}^{-1}\boldsymbol{A})\sum_{j=0}^{n-1}(\Delta \boldsymbol{R}_{t_j}) - (\Delta t)^2(\boldsymbol{A}'\boldsymbol{\Sigma}^{-1}\boldsymbol{A})\sum_{j=0}^{n-1}(\boldsymbol{R}^* - \boldsymbol{R}_{t_j}), \qquad (25)$$

setting equal to zero and solving for $\boldsymbol{R}^*$, we obtain

$$(\Delta t)\left(\boldsymbol{A}'\boldsymbol{\Sigma}^{-1}\right)\left(\boldsymbol{R}_{t_n} - \boldsymbol{R}^0\right) = (\Delta t)^2\left(\boldsymbol{A}'\boldsymbol{\Sigma}^{-1}\boldsymbol{A}\right)\sum_{j=0}^{n-1}\left(\boldsymbol{R}^* - \boldsymbol{R}_{t_j}\right), \qquad (26)$$

where we computed the telescope sum $\sum_{j=0}^{n-1}(\Delta \boldsymbol{R}_{t_j}) = \boldsymbol{R}_{t_n} - \boldsymbol{R}^0 \equiv \boldsymbol{R}_T - \boldsymbol{R}^0$.

Assuming both $\boldsymbol{A}$ and $\boldsymbol{\Sigma}$ are non-singular, multiplying with $(\boldsymbol{A}'\boldsymbol{\Sigma}^{-1})^{-1}$ from the left, we can solve for $\boldsymbol{R}^*$ to obtain

$$\boldsymbol{R}^* = \frac{1}{n}\sum_{j=0}^{n-1}\boldsymbol{R}_{t_j} + \frac{1}{T}\boldsymbol{A}^{-1}\left(\boldsymbol{R}_{t_n} - \boldsymbol{R}^0\right) \text{ where } T = n(\Delta t). \qquad (27)$$

Differentiating with respect to $\boldsymbol{A}$, we get

$$\frac{\partial}{\partial \boldsymbol{A}}\ell(\boldsymbol{\theta}|\boldsymbol{R}) = (\Delta t)\frac{\partial}{\partial \boldsymbol{A}}\sum_{j=0}^{n-1}(\boldsymbol{\Sigma}^{-1}\Delta\boldsymbol{R}_{t_j})'\boldsymbol{A}(\boldsymbol{R}^* - \boldsymbol{R}_{t_j}) \tag{28}$$

$$-\frac{(\Delta t)^2}{2}\frac{\partial}{\partial \boldsymbol{A}}\sum_{j=0}^{n-1}(\boldsymbol{A}(\boldsymbol{R}^* - \boldsymbol{R}_{t_j}))'\boldsymbol{\Sigma}^{-1}(\boldsymbol{A}(\boldsymbol{R}^* - \boldsymbol{R}_{t_j})) \tag{29}$$

$$= (\Delta t)\sum_{j=0}^{n-1}\boldsymbol{\Sigma}^{-1}(\boldsymbol{R}_{t_{j+1}} - \boldsymbol{R}_{t_j})(\boldsymbol{R}^* - \boldsymbol{R}_{t_j})' \tag{30}$$

$$-(\Delta t)^2\sum_{j=0}^{n-1}\boldsymbol{\Sigma}^{-1}\boldsymbol{A}(\boldsymbol{R}^* - \boldsymbol{R}_{t_j})(\boldsymbol{R}^* - \boldsymbol{R}_{t_j})'. \tag{31}$$

where we invoked Supplemental Lemma S2.2. Setting the partial derivative equal to zero and solving for $\boldsymbol{A}$, we get

$$(\Delta t)\boldsymbol{\Sigma}^{-1}\sum_{j=0}^{n-1}(\Delta\boldsymbol{R}_{t_j})(\boldsymbol{R}^* - \boldsymbol{R}_{t_j})' = (\Delta t)^2\boldsymbol{\Sigma}^{-1}\boldsymbol{A}\sum_{j=0}^{n-1}(\boldsymbol{R}^* - \boldsymbol{R}_{t_j})(\boldsymbol{R}^* - \boldsymbol{R}_{t_j}).' \tag{32}$$

Multiplying from the left with $\boldsymbol{\Sigma}$ and solving for $\boldsymbol{A}$, we obtain

$$\boldsymbol{A} = \frac{1}{\Delta t}\left(\sum_{j=0}^{n-1}(\boldsymbol{R}_{t_{j+1}} - \boldsymbol{R}_{t_j})(\boldsymbol{R}^* - \boldsymbol{R}_{t_j})'\right)\left(\sum_{j=0}^{n-1}(\boldsymbol{R}^* - \boldsymbol{R}_{t_j})(\boldsymbol{R}^* - \boldsymbol{R}_{t_j})'\right)^{-1}. \tag{33}$$

Lastly, we differentiate with respect to $\boldsymbol{\Sigma}$ to obtain

$$\frac{\partial}{\partial \boldsymbol{\Sigma}}\ell(\boldsymbol{\theta}|\boldsymbol{R}) = -\frac{T}{2}\frac{\partial}{\partial \boldsymbol{\Sigma}}\log|\boldsymbol{\Sigma}| \tag{34}$$

$$-\frac{1}{2}\frac{\partial}{\partial \boldsymbol{\Sigma}}\sum_{j=0}^{n-1}\left\langle\left(\Delta\boldsymbol{R}_{t_j} - \boldsymbol{A}(\boldsymbol{R}^* - \boldsymbol{R}_{t_j})(\Delta t)\right)\left(\Delta\boldsymbol{R}_{t_j} - \boldsymbol{A}(\boldsymbol{R}^* - \boldsymbol{R}_{t_j})(\Delta t)\right)', \boldsymbol{\Sigma}^{-1}\right\rangle_{\mathcal{F}}.$$

By employing Supplemental Lemmas S2.3 and S2.4, we simplify to find

$$\frac{\partial}{\partial \boldsymbol{\Sigma}}\ell(\boldsymbol{\theta}|\boldsymbol{R}) = -\frac{T}{2}\boldsymbol{\Sigma}^{-1} \tag{35}$$

$$+\frac{1}{2}\boldsymbol{\Sigma}^{-1}\sum_{j=0}^{n-1}\left(\Delta\boldsymbol{R}_{t_j} - \boldsymbol{A}(\boldsymbol{R}^* - \boldsymbol{R}_{t_j})(\Delta t)\right)\left(\Delta\boldsymbol{R}_{t_j} - \boldsymbol{A}(\boldsymbol{R}^* - \boldsymbol{R}_{t_j})(\Delta t)\right)'\boldsymbol{\Sigma}^{-1}.$$

Setting the latter expression equal to zero and multiplying from the left with $\boldsymbol{\Sigma}$ from the left and from the right, we obtain:

$$\boldsymbol{\Sigma} = \frac{1}{T}\sum_{j=0}^{n-1}\left(\Delta\boldsymbol{R}_{t_j} - \boldsymbol{A}(\boldsymbol{R}^* - \boldsymbol{R}_{t_j})(\Delta t)\right)\left(\Delta\boldsymbol{R}_{t_j} - \boldsymbol{A}(\boldsymbol{R}^* - \boldsymbol{R}_{t_j})(\Delta t)\right)'. \tag{36}$$

Combining Eqs. (27), (33) and (36), the claim follows.

References

1. Aït-Sahalia, Y.: Nonparametric pricing of interest rate derivative securities. In: NBER Working Paper Series, 5345 (1995)
2. Aït-Sahalia, Y., Kimmel, R.L.: Estimating affine multifactor term structure models using closed-form likelihood expansions. J. Financ. Econ. **98**(1), 113–144 (2010)
3. European Central Bank. Euro area yield curves. https://www.ecb.europa.eu/stats/financial_markets_and_interest_rates/euro_area_yield_curves/html/index.en.html. Accessed 05 Feb 2023
4. Basu, A., Harris, I.R., Hjort, N.L., Jones, M.C.: Robust and efficient estimation by minimising a density power divergence. Biometrika **85**(3), 549–559 (1998)
5. Bian, L., Li, Z., Yao, H.: Time-consistent strategy for a multi-period mean-variance asset-liability management problem with stochastic interest rate. J. Industr. Manage. Optim. **17**(3), 1383–1410 (2021)
6. Black, F.: Capital market equilibrium with restricted borrowing. J. Bus. **45**(3), 444–455 (1972)
7. Bu, R., Giet, L., Hadri, K., Lubrano, M.: Modeling multivariate interest rates using time-varying copulas and reducible nonlinear stochastic differential equations. J. Financ. Economet. **9**(1), 198–236 (2011)
8. Cox, J.C., Ingersoll, J.E., Jr., Ross, S.A.: A theory of the term structure of interest rates. Econometrica **53**(2), 385–407 (1985)
9. Dai, Q., Singleton, K.J.: Specification Analysis of affine term structure models. J. Financ. **55**(5), 1943–1978 (2000)
10. Duffie, D., Glynn, P.: Estimation of continuous-time Markov processes sampled at random time intervals. Econometrica **72**(6), 1773–1808 (1997)
11. Egorov, A.V., Li, H., Ng, D.: A tale of two yield curves: Modeling the joint term structure of dollar and euro interest rates. J. Econom. **162**(1), 55–70 (2011)
12. Eraker, B.: MCMC analysis of diffusion models with application to finance. J. Bus. Econ. Stat. **19**(2), 177–191 (2001)
13. Gallant, A.R., Tauchen, G.: Which moments to match? Economet. Theor. **12**(4), 657–681 (1996)
14. Gourieroux, C., Monfort, A., Renault, E.: Indirect inference. J. Appl. Economet. **8**(S1), S85–S118 (1993)
15. Hamilton, J.D.: Time Series Analysis. Princeton University Press, Princeton (1994)
16. Huang, Z., Jiang, T., Wang, Z.: On a multiple credit rating migration model with stochastic interest rate. Math. Methods Appl. Sci. **43**(12), 7106–7134 (2020)
17. Huber, P.J., Ronchetti, E.M.: Robust Statistics. Wiley Series in Probability and Statistics, vol. 52. Wiley, New York (2009)
18. Hull, J., White, A.: Pricing interest-rate-derivative securities. Rev. Financ. Stud. **3**(4), 573–592 (1990)
19. Kessler, M., Sørensen, M.: Estimating equations based on eigenfunctions for a discretely observed diffusion process. Bernoulli **5**(2), 299–314 (1999)
20. Mao, H., Wen, Z.: Optimal decision on dynamic insurance price and investment portfolio of an insurer with multi-dimensional time-varying correlation. J. Quant. Econ. **18**(1), 29–51 (2020)
21. Markowitz, H.: The utility of wealth. J. Polit. Econ. **60**(2), 151–158 (1952)
22. Mehrdoust, F., Najafi, A.R.: A short memory version of the Vasicek model and evaluating European options on zero-coupon bonds. J. Comput. Appl. Math. **375**, 112796 (2020)

23. Merton, R.C.: On estimating the expected return on the market: an exploratory investigation. J. Financ. Econ. **8**(4), 323–361 (1980)
24. Pokojovy, M., Jobe, J.M.: A robust deterministic affine-equivariant algorithm for multivariate location and scatter. Comput. Stat. Data Anal. **172**(107475) (2022)
25. Rousseeuw, P., Van, A.S., Van, D.K., Agulló, J.: Robust multivariate regression. Technometrics **46**, 293–305 (2022)
26. Rousseeuw, P.J., Leroy, A.: Robust Regression and Outlier Detection. Wiley, New York (1987)
27. Shen, G.J., Wang, Q.B., Yin, X.W.: Parameter estimation for the discretely observed Vasicek model with small fractional Lévy noise. Acta Math. Sinica Engl. Ser. **36**(4), 443–461 (2020)
28. Stanton, R.: A nonparametric model of term structure dynamics and the market price of interest rate risk. J. Financ. **52**(5), 1973–2002 (1997)
29. US Department of the Treasury. Daily treasury yield curve rates (2023). https://www.treasury.gov/resource-center/data-chart-center/interest-rates/Pages/TextView.aspx?data=yield. Accessed 05 Feb 2023
30. Vasicek, O.: An equilibrium characterization of the term structure. J. Financ. Econ. **5**(2), 177–188 (1977)
31. Wang, X., Phillips, P.C.B., Yu, J.: Bias in estimating multivariate and univariate diffusions. J. Econom. **161**(2), 228–245 (2011)

Stochastic Bike-Sharing Transport Network Design

Gang Wang[1], Yiwei Fan[2], and Xiaoling Lu[2(✉)]

[1] University of Massachusetts Dartmouth, North Dartmouth, MA 02747, USA
[2] Renmin University of China, Beijing 100872, China
xiaolinglu@ruc.edu.cn

Abstract. As a transport mode, bike-sharing has gained popularity worldwide because it is environmentally friendly and cost-efficient. However, as a bike-sharing network grows, operating costs at rental centers increase. The problem is determining the locations of rental centers to open and the number of bicycles that will be transferred daily between rental centers while minimizing the total operating costs. We present a stochastic programming model and a Benders decomposition-based hybrid algorithm. We consider two scenarios for demand-return machine learning models - time series-based prediction and weather-based forecasting. Finally, we provide a case study of developing a bike-sharing network in New York City to verify the significance of the proposed models. We also evaluate the performances of demand-return prediction models and the impact of the relative ratio between demand and return on bike-sharing network design. We find no bicycle transfer if the penalty cost for a rental station has an inverse linear relationship with the ratio of returns to rentals. Nevertheless, when the penalty cost is exponentially dependent on the negative ratio of returns to rentals, bicycle transfer occurs between rental stations with large ratios.

Keywords: Bike sharing transport network · Uncertain demand and return · Stochastic programming · Machine learning · Benders decomposition

1 Introduction

Urban cities worldwide are increasingly implementing bike-sharing programs, reducing air pollution and pressure on traffic congestion. Bike-sharing programs offer bicycles scattered throughout a city for public use at low cost and high convenience by allowing users to pick up a bike at any self-serve bicycle center and return it to another. As of June 2014, public bike-sharing systems were available in 50 countries on five continents, including 712 cities, operating approximately 806,200 bicycles at 37,500 stations. Since New York City launched its Citi Bike program, bicycle centers have doubled from 330 to 674 between August 2015 and November 2016. As bike-sharing networks grow, operating costs increase

© The Author(s), under exclusive license to Springer Nature Switzerland AG 2024
H. Han and E. Baker (Eds.): SDSC 2023, CCIS 2113, pp. 19–33, 2024.
https://doi.org/10.1007/978-3-031-61816-1_2

significantly, while bicycle availability at bicycle centers decreases. The keys to achieving a successful bike-sharing program are reaching accordance between the locations of bike-sharing centers, and the demand Lin and Yang (2011) and the ability to deal with varying user demand. The purpose of this paper is to combine machine learning and stochastic optimization to design an efficient and effective bicycle-sharing network with uncertain demand.

Bicycle sharing systems have undergone rapid growth in recent years as an urban transportation mode due to advances in tracking technology. Most research in the literature focuses on adopting optimization techniques for bike distribution and routing among rental stations. Forma et al. (2015) introduce a three-step optimization-based heuristic to solve a bike repositioning problem with the instances of up to 200 stations and three vehicles. Li et al. (2016) suggest a combined hybrid genetic algorithm for a new static bicycle repositioning problem with multiple types of bikes. Dell'Amico et al. (2016) develop a destroy-and-repair metaheuristic algorithm to study a bike redistributing problem among the stations of a bike-sharing system using a fleet of capacitated vehicles. Furthermore, Kadri et al. (2016) use a branch-and-bound algorithm to solve a vehicle scheduling problem in bike-sharing systems to minimize operating costs at rental stations in less waiting time. Cruz et al. (2017) offer an iterated local search-based heuristic for solving a static bike-rebalancing problem among stations in self-service bike-sharing systems. After the test on 980 benchmark instances in the literature, their algorithm obtains better results than existing methods. Ghosh et al. (2017) recommend two optimization approaches for a bike-repositioning and vehicle routing problem with future demand. Ho and Szeto (2017) develop a hybrid neighborhood search to examine a bike-repositioning problem, including a pick-up and delivery vehicle routing problem in bike-sharing systems. Szeto and Shui (2018) utilize an enhanced artificial bee colony algorithm to address a bike-repositioning problem by minimizing the deviation between demand disappointment and service time. However, few studies discuss bicycle-sharing network design and repositioning problems with stochastic demand using three pillars of business analytics (i.e., descriptive, prescriptive, and predictive analytics). Lin and Yang (2011) study the design problem of the bicycle sharing system by solving a mixed-integer nonlinear program with a commercial solver. Larsen et al. (2013) use multi-criteria methods and a GIS methodology to obtain optimal locations with new routes in the Montreal cycling network. Yan et al. (2017) consider locating bike rental stations, scheduling bike fleet, and routing with deterministic and stochastic demands by creating a time-space network model. They used CPLEX to solve deterministic models directly, while they created a threshold-accepting-based heuristic for stochastic models. However, the papers above reviewed miss stochastic demand or demand forecasting or both, although they consider bike-sharing network design problems. Moreover, Vogel (2016) integrates demand data analysis and mathematical optimization for bike-sharing network design and bike relocation. However, this paper neglects the location problem of bike rental stations. Gao (2019) only consider rental demand forecast in a bike-sharing system by developing a moment-based predictive model. To the best of our knowledge, little research in the literature uses the combination of descriptive, prescriptive, and predictive analytics to investigate bike-sharing network design with stochastic demand.

Therefore, in this paper, we formulate a two-stage stochastic programming model that incorporates the three decisions mentioned above. Furthermore, we consider two scenarios for demand-return forecasting: one with time series-based prediction and the other concerning weather-based prediction. We also present a Benders decomposition-based algorithm that uses demand-return forecast results as input to solve this problem. In addition, we provide a real-life case study regarding the Citi Bike program in New York City to validate our proposed models and solution approaches. Finally, we evaluate the performances of demand-return prediction models and the impact of the relative ratio between demand and return on bike-sharing network design.

The rest of this paper is organized as follows. Section 2 proposes a two-stage stochastic programming model for bicycle-sharing network design with stochastic demand. Section 3 develops a solution approach based on the Benders decomposition. Section 4 presents computational results and managerial insights. Finally, Sect. 5 presents our conclusions.

2   Model Formulation

2.1   Stochastic Programming Model

Consider a transport network $\mathcal{G} = (\mathcal{N}, \mathcal{A})$, where $\mathcal{N}$ denotes the set of nodes and $\mathcal{A}$ is the set of arcs. The nodes represent bicycle rental centers located in different areas of a city, while arcs represent possible transfers of bicycles between rental centers. The operating expenses of opening a rental center $i \in \mathcal{N}$ are f_i. If opened, each rental center $i \in \mathcal{N}$ faces stochastic customer demand d_i and customer return r_i. Bicycle transfer between rental centers incurs a transfer cost v_{ij} per unit from i to j, $i, j \in \mathcal{N}$. We assume that bicycle transfer occurs between rental centers once each day and that the number s_i of bicycles that are available at center i is known before the transfer occurs. Since each rental center $i \in \mathcal{N}$ offers a limited number of bicycles, i.e., its capacity c_i, bicycle stockout z_i incurs penalty cost h_i per unit. The problem involves three types of decisions: configuration decisions, e.g., which bicycle rental centers to open; demand forecasting decisions; and operational decisions, e.g., the number of bicycles that are transferred from one center to another. Let us denote by y_i configuration decisions, $y_i = 1$ if rental center $i \in \mathcal{N}$ is open, and 0 otherwise. Also, let x_{ij} be the amount of transferred bicycles from i to j, $(i, j) \in \mathcal{A}$. We summarize the notations as follows.

Sets and Indices
   $\mathcal{N}$ = set of bicycle rental centers (RC), i.e., $i \in \mathcal{N}$
   $\mathcal{A}$ = set of possible transfer between RCs
Parameters
   f_i = fixed operating cost (\$) at RC_i;

d_i = random demand at RC_i in a day;
r_i = random return at RC_i in a day;
v_{ij} = transfer cost per unit from i to j;
s_i = the number of vehicles available at center i before transfer occurs;
c_i = capacity of center i;
h_i = penalty per unit for unmet demand at center i.

Decision Variables

$$y_i = \begin{cases} 1, & \text{if center } i \text{ is open} \\ 0, & \text{otherwise} \end{cases}$$

x_{ij} = the number of vehicles transferred from i to j
z_i = the number of vehicles for failing to meet demand at center i

Since the objective of the problem is to minimize the total operating expenses of opening bicycle rental centers, a stochastic integer program formulates the stochastic rental network design problem as follows:

$$P: \quad \min \quad \sum_{i \in \mathcal{N}} f_i y_i + E[g(y_i, d_i, r_i)] \tag{1}$$

$$\text{s.t.} \quad y_i \in \{0, 1\}, i \in \mathcal{N}$$

where $E[g(y, d, r)] = \sum_{(i,j) \in \mathcal{A}} v_{ij} x_{ij} + \sum_{i \in \mathcal{N}} h_i z_i$ is the optimal value of the following problem

$$\min \quad \sum_{(i,j) \in \mathcal{A}} v_{ij} x_{ij} + \sum_{i \in \mathcal{N}} h_i z_i \tag{2}$$

s.t.

$$s_i y_i - \sum_j x_{ij} + \sum_j x_{ji} + z_i \geq d_i, \quad i \in \mathcal{N} \tag{3}$$

$$(s_i + r_i) y_i - \sum_j x_{ij} + \sum_j x_{ji} \leq c_i y_i, \quad i \in \mathcal{N} \tag{4}$$

$$\sum_j x_{ij} \leq s_i y_i, \quad i \in \mathcal{N} \tag{5}$$

$$y_i \in \{0, 1\}; x_{ij}, z_i \geq 0 \text{ and integer}, i, j \in \mathcal{N} \tag{6}$$

The objective function (1) minimizes the total costs, including the costs of opening rental centers, transferring bicycles between rental centers, and penalizing unmet bicycle demand. Constraints (3) require that the number of bicycles at center i satisfy customer demand. Constraints (4) state that the number of bicycles at center i can not exceed its capacity. Constraints (5) stipulate that the number of transferred bicycles from center i can not be greater than the number of available bicycles. Constraints (6) enforce nonnegativity and integrality. Throughout, we shall denote by P the studied problem.

3   Solution Approach

In this section, we give a matrix form of P and present a hybrid algorithm for solving P based on Benders decomposition and the prediction models defined in Sect. 2.

For each pair d and r, we introduce a matrix form of P as follows.

$$\min f^T y + v^T x + h^T z$$

$$s.t.$$

$$-sy + Ax - z \leq -d,$$
$$(s + r - c)y - Ax \leq 0,$$
$$-sy + Bx \leq 0,$$
$$x, z \geq 0 \text{ and integer,}$$

where f, v, h, d, r, s, c are the vectors that correspond to the costs of operating rental centers, transfer, penalty, demands, and returns, the number of available bicycles and the capacity of the rental centers, respectively. A and B are the coefficient matrices in Constraints (3), (4), and (5).

To rewrite the above model for a period of time, namely, m days, we decide whether a rental station i should be open or closed, that is, $y_i = 1$ or 0. Using the average transfer costs in the objective function, the original problem can be transformed into the following:

$$\min_{y} f^T y + G(y) \tag{7}$$

$$s.t. \tag{8}$$

$$y \in \{0, 1\}^{|\mathcal{N}|} \tag{9}$$

where $G(y) = \frac{1}{m} \sum_{k=1}^{m} g_k(y, d^{(k)}, r^{(k)}, s^{(k)})$ is the optimal value of the problem below:

$$\min v^T x + h^T z \tag{10}$$

$$s.t. \tag{11}$$

$$Ax - z \leq s^{(k)} y - d^{(k)} \tag{12}$$

$$-Ax \leq -(s^{(k)} + r^{(k)} - c)y \tag{13}$$

$$Bx \leq s^{(k)} y \tag{14}$$

$$x \in \mathbb{Z}^{|\mathcal{N}| \times |\mathcal{N}|} \tag{15}$$

$$z \in \mathbb{Z}^{|\mathcal{N}|} \tag{16}$$

To solve P, we present a hybrid algorithm based on forecasting models and the Benders decomposition developed by Santoso et al. (2005). The algorithm first predicts bicycle demand and returns over the next m days, which are the input to the Benders decomposition, using the forecasting models in Sect. 2. The algorithm, denoted by Algorithm BA, appears in the following.

Algorithm 1. Algorithm BA

1: Step 0: Set $j = 0$ and upper bounds $ub^0 = +\infty$ and ϵ be the pre-specified error tolerance.

2: Step 1: Predict the bicycle demand $d^{(k)}$ and returns $r^{(k)}$ over m days.

3: Step 2: Solve the master problem MP below:

$$\min_{y,\theta} f^T y + \theta \tag{17}$$

$$s.t. \quad y \in \{0,1\}^{|\mathcal{N}|} \tag{18}$$

$$\theta \geq \frac{1}{m} \sum_{k=1}^{m} F_k^l(y, d^{(k)}, r^{(k)}, s^{(k)}), l = 1, \cdots, j \tag{19}$$

If MP is infeasible, stop. If MP is unbounded, choose any (y^j, θ^j) with $f^T y^j + \theta^j < ub^j$. Whenever (y^j, θ^j) exists, go to Step 3.

4: Step 3: Given y^j in Step 2, solve the problem (10) in the second stage. Denote the corresponding optimal dual functions by $F_k^{j+1}(y^j, d^{(k)}, r^{(k)}, s^{(k)})$ for $k = 1, \cdots, m$. Let $\hat{F}_k^{j+1}(y^j, d^{(k)}, r^{(k)}, s^{(k)})$ denote the optimal value of dual problem of (10). Set $ub^{j+1} = \min\{ub^j, f^T y^j + \frac{1}{m} \sum_{k=1}^{m} \hat{F}_k^{j+1}(y^j, d^{(k)}, r^{(k)}, s^{(k)})\}$. If $ub^{j+1} - lb^j < \epsilon$, then y^j is optimal and ub^{j+1} is the optimal value of P; else, go to Step 4.

5: Step 4: Add $\theta \geq \frac{1}{m} \sum_{k=1}^{m} F_k^{j+1}(y, d^{(k)}, r^{(k)}, s^{(k)})$ to MP where $F_k^{j+1}(y, d^{(k)}, r^{(k)}, s^{(k)}) = p_{j+1}^T y + q_{j+1}$. Note that

$$p_{j+1}^T = ((\zeta_j^{(k)} \circ s^{(k)})^T - (\eta_j^{(k)} \circ (s^{(k)} + r^{(k)} - c))^T$$
$$+ (\mu_j^{(k)} \circ s^{(k)})^T)$$
$$q_{j+1} = -(\zeta_j^{(k)})^T d^{(k)}$$

where $(\zeta_j^{(k)}, \eta_j^{(k)}, \mu_j^{(k)})$ is the optimal solution to the dual problem of (10). To improve the algorithmic convergence, set $y_i = 1$ if $\bar{d}_i \geq d_0$ or $\bar{r}_i \geq r_0$, where $\bar{d}_i = \frac{1}{m} \sum_{k=1}^{m} d_i^{(k)}$, $\bar{r}_i = \frac{1}{m} \sum_{k=1}^{m} r_i^{(k)}$, and d_0, r_0 is given. Set $j := j + 1$ and go to Step 1.

---

4 Computational Results and Discussion

In this section, we first describe the public bicycle system in New York City and apply clustering techniques from machine learning to divide areas so that bicycle transfers occur within each area. Then we forecast demand and return using the predictive models proposed in Sect. 3. Finally, we solve this real case through

Algorithm BA and obtain managerial insights by considering different relative ratios between demand and return.

4.1   Data Description and Clustering

We have collected data on the Public Bicycle System of New York City from the website of the Citi Bike program (Citibike, 2017). The Citi Bike program offers monthly transaction data for 674 rental centers between July 1, 2013, to February 28, 2017. Each record includes the start time, start station, stop station, station id, station name, station longitude, and station latitude. After processing the data, we deleted seven rental centers listed in Table 1. Of these seven rental centers, the first two have no latitude and longitude information; the next three are far away from other rental centers and can be considered outliers, and the last two have the same latitude and longitude values but different ids and names.

Table 1. Rental centers with abnormal location information

Station ID	Station Name	Station Latitude	Station Longitude
3240	NYCBS Depot BAL - DYR	0	0
3446	NYCBS Depot - STY - Valet Scan	0	0
160	E 37 St & Lexington Ave	40.44535	−73.9782
255	NYCBS Depot - SSP	40.64677	−74.0165
3239	Bressler	40.64654	−74.0166
3245	Kiosk in a box Motivate	40.64668	−74.0163
3040	SSP Tech Workshop	40.64668	−74.0163

Figure 1 shows the number of active rental centers each month from July 2013 to February 2017. As shown in Fig. 1, the number of rental centers remains approximately 330 between July 2013 and August 2015, while the number of rental centers doubles after one and a half years. Because of the Citi Bike program, the demand for bicycles has significantly increased in New York City. Figure 2 shows many new rental centers established since August 2015, which are denoted by diamonds and located at the periphery, and the rental centers built before 2015, which are denoted by triangles and located in the center. Moreover, some rental centers are pretty busy and have a high probability of stockouts, while others are often idle, with the low demand and return, which results in significant operational inefficiencies. Therefore, it is necessary to redesign the public bike-sharing network in New York City. In actual operations, bicycles move between adjacent rental centers rather than distant ones. Hence, we partition the 667 rental centers into 20 clusters, each including 20 to 60 rental centers, using the K-means clustering technique from machine learning and each rental center's latitude and longitude information. Citi Bike can determine each cluster based on the actual situation. Figure 3 displays the clustering results with the number of rental centers in each group.

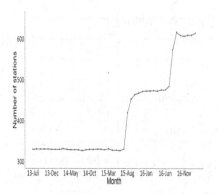

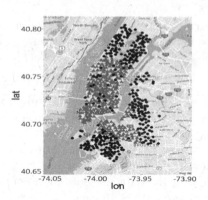

Fig. 1. The number of active stations in months

Fig. 2. Stations established before (triangle) and after August 2015 (diamond)

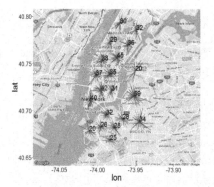

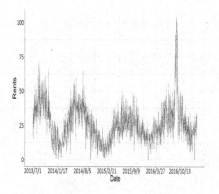

Fig. 3. The clusters of rental centers in NYC

Fig. 4. The demand time series demand of the 82th rental center

4.2 Demand-Return Analysis

Data Preparation and Forecasting. This section applies statistical models to forecast the bicycle demand and return for each rental center. We first calculate the number of bicycles needed each day and the returns between 6:00 am to 12:00 pm for each rental center. The start dates are different because stations are not built and used simultaneously. Therefore, for each station, we consider as the training set the records from the first day in which there are rentals or returns to Feb 21, 2017, and forecast the numbers of rentals and returns from Feb 22, 2017, to Feb 28, 2017 (the last week of the study period). The process for forecasting the rentals and returns for each station appears below.

– For a station, if the length of the training set is less than 14 days,
  • *Situation 1:* if the length of the training set is less than seven days, then we use the average value of the training record as the forecast from Feb 22, 2017, to Feb 28, 2017. There are six sites in this group.

- **Situation 2:** if the length of the training set is less than 14 days and greater than seven days, we directly select the record from Feb 15, 2017, to Feb 21, 2017, as the forecast from Feb 22, 2017, to Feb 28, 2017. There is one site in this group.
- For a station, if the length of the training set is greater than 14 days,
  - **Situation 3:** if no bicycle was demanded or returned in the last week of the training set from Feb 15, 2017, to Feb 21, 2017, we assume this situation will continue into the next week, that is, this site may be closed or undergoing maintenance. Therefore, the prediction will be zero from Feb 22, 2017, to Feb 28, 2017. There are 63 such sites. In practice, the Citi Bike company can decide this situation according to their daily operations, whereas here we make this assumption for simplicity,
  - **Situation 4:** For the remaining 597 sites, we use exponential smoothing and random forest methods for prediction.
    * **Situation 4.1:** The demand and return for some sites in a certain period are always zero. If the number of such days is greater than or equal to 7, we consider these records abnormal (the site may be closed for maintenance). We use basic exponential smoothing for the time series model to interpolate these data to ensure continuity in time. For random forests, we discard these records. There are 127 such sites;
    * **Situation 4.2:** There are four days, from January 23th, 2016, to Jan 26, 2016, in which there is no demand or return for all 597 sites. Therefore, we also interpolate these data for time-series prediction and discard them for random forests.

Figure 4 shows a scatter diagram of the bicycle demand at rental center 82 (St. James Pl. & Pearl St.). The time series shows the trend and seasonality. For example, in winter (January to March of each year), bicycle demand was relatively low, at approximately 5. In contrast, bicycle demand was high, at approximately 30, in summer (July to October of each year), especially in October 2016, in which it increased to over 75. We use the aforementioned exponential smoothing algorithms to predict each station's bicycle demand and returns in Situation 4.

We wrote a crawler code for the random forest to obtain weather data for the New York area from the website https://www.wunderground.com/ from July 2013 to Feb 28, 2017, including temperature, visibility, and wind speed, precipitation, and other variables (20 in all). In addition, we consider the year, month, week, holidays, and demand or return in the previous seven days as predictors. In forecasting, we perform a one-step forward forecasting procedure. First, we use the predicted values for the current day as the input to forecast the values for the next day. Then, we forecast the bicycle demand from Feb 22, 2017, to Feb 28, 2017.

Figure 5 shows the predicted bicycle demand versus the actual demand for two stations, namely, 82 and 116. Station 82 is relatively small, and the bicycle demand is approximately 30. On Feb 22, 2017, the actual demand was 30. The

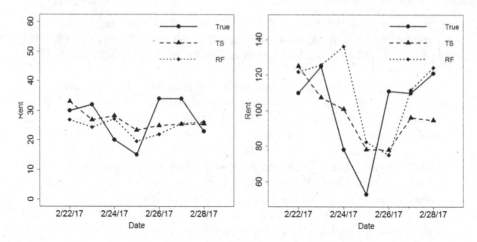

Fig. 5. Comparison of actual and forecast values of demand at stations 82 (left) and 116 (right) between time series and random forest models

forecasted demands by time series and random forests are 33 and 27, respectively. Station 116 is relatively large, the demand is approximately 100, and the actual demand was 110 on Feb 22, 2017. The forecasted demands by time series and random forests are 125 and 122, respectively. The same observations apply equally to the other rental stations, so they are not discussed here.

Comparison Between Two Statistical Models. Let us compare the predictions between two statistical models. We first introduce some criteria, which are listed in Table 2, where r_i denotes the true value and $\hat{r}_i$ is the predicted value.

Table 3 shows the median values of these statistics for 597 rental centers and the percentage of centers for which random forest performs well. TS stands for the exponential smoothing method, and RF stands for the random forest method. The median values of MAE are approximately 9 for both models, which means that, on average, the difference between the actual value and predicted value is 9. The multi-order exponential smoothing method better forecasts bicycle demand, while the random forest is better for bicycle return forecasting.

Table 2. The measures of forecasting accuracy

	Explanation	Calculation equation
MAE	Mean absolute error	$(1/7)\sum_{i=1}^{7} |r_i - \hat{r}_i|$
RMSE	Root mean squared error	$\sqrt{(1/7)\sum_{i=1}^{7} (r_i - \hat{r}_i)^2}$
MAPE	Mean absolute percentage error	$(1/7)\sum_{i=1}^{7} |(r_i - \hat{r}_i)/r_i|$
TIC	Theil's Inequality Coefficient	$\dfrac{\sqrt{(1/7)\sum_{i=1}^{7}(r_i-\hat{r}_i)^2}}{(\sqrt{(1/7)\sum_{i=1}^{7}\hat{r}_i^2}+\sqrt{(1/7)\sum_{i=1}^{7}r_i^2})}$

Table 3. Comparison of demand forecasts between time series and random forecast models

	Demand			Return		
	TS	RF	RF better (%)	TS	RF	RF better (%)
MAE	9.54	9.74	36.35	8.94	8.25	52.76
RMSE	11.24	11.90	36.35	10.85	10.36	51.59
MAPE	0.24	0.24	45.90	0.22	0.21	54.10
TIC	0.14	0.15	33.17	0.14	0.14	50.25

4.3   Network Planning Analysis

Parameter Settings. We first specify how to define the values of parameters such as fixed operating cost, transfer cost, and penalty cost. Note that v_{ij} is the transfer cost per unit from center i to center j. Suppose that it is proportional to the distance between i and j, denoted as $dist_{ij}$ kilometers. Assume that the cost to transfer one vehicle one kilometer is 0.2. The fixed cost f_i is 2 for all i. In practice, the Citi Bike company can set s_i and c_i according to the administrative operations. Here we generate their values by simulation. Assuming there are K vehicles in total, for each site, the value of supply s_i is evenly distributed from $0.8 * Kd_i / \sum_i d_i$ to $1.2 * Kd_i / \sum_i d_i$ where d_i represents prediction for bicycle demand and we set $K = 10000$ according to the official website (https://www. citibikenyc.com/). We assume that the value of capacity c_i is large enough; there is no constraint regarding capacity. We then consider the penalty cost $h_i, i \in \mathcal{N}$.

Our model did not consider the relationship between demand and return. However, part of the demand can be satisfied with the return. To consider the impact of the number of returns, we introduce the ratio r_i/d_i, which means that there are r_i/d_i returns when one bicycle is rented. Thus, the penalty h_i can be expressed as a function of r_i/d_i and has the following properties:

1. $h_i \geq 0$;
2. h_i is a decreasing function of r_i/d_i;
3. $h_i = h_0 > 0$ if $r_i = 0$, that is, if there is no return, h_i will be a positive constant;
4. $h_i = 0$ if $d_i = 0$.

For simplicity, we set $h_0 = 1$ as a baseline. Then, we give two forms of h_i and explain their meanings.

The first one is

$$h_i^{(1)} = \begin{cases} (1 - \frac{r_i}{d_i})_+ & \text{if } d_i \neq 0 \\ 0 & \text{if } d_i = 0 \end{cases}$$

Here, $u_+ = u$ if $u > 0$; otherwise, $u_+ = 0$. Obviously, it satisfies the properties that are stated above. When $r_i \geq d_i$, $h_i^{(1)} = 0$. This means that if the number of returns is larger than the demand at some site, then the demand can be satisfied after a period of waiting time. When $r_i < d_i$, $h_i^{(1)} = 1 - r_i/d_i$. Then, the penalty term in the objective function (2) is $z_i - z_i(r_i/d_i)$. r_i/d_i means there are r_i/d_i returns when one bicycle is rented. Thus, for a shortage of z_i bicycles, there are $z_i(r_i/d_i)$ returns, which can offset part of the shortage. Finally, only $z_i - z_i(r_i/d_i)$ rental requests cannot be satisfied.

However, no penalty is incurred when $r_i > d_i$. We assume that $h_i > 0$ always holds and h_i is a strictly monotone decreasing function of r_i/d_i. Hence, the second form is given as:

$$h_i^{(2)} = \begin{cases} exp\{-\frac{r_i}{d_i}\} & \text{if } d_i \neq 0, \\ 0 & \text{if} d_i = 0 \end{cases}$$

Note that $exp\{-r_i/d_i\}$ is a strictly monotone decreasing function of r_i/d_i, which is an improvement compared to $h_i^{(1)}$. In the following, we will present the transfer plan for different $h_i^{(1)}$ and $h_i^{(2)}$.

Results. Table 4 shows the results regarding which stations should be closed in the next week across all clusters. The solution for both penalties, namely, $h_i^{(1)}$ and $h_i^{(2)}$, are same. For example, for cluster 11, there are 25 stations, of which 2 will be closed, namely those with ids 3401 and 3432.

Although the results for y for both penalties are the same, there are differences between the transfer plans. Figure 6 shows the transfer plans for the first day, namely, Feb 22, 2017, for group 11. Here, we use the number of points to represent the number of bicycles rented in each site, and the value along the edge shows the transfer number. Note that the 20th site, with station id. 3401, and the 25th site, with station id. Three thousand four hundred thirty-two will be closed in our solution, and we represent them as squares.

For example, in the 3rd station, the prediction for demand is 24 and for return is 23, so $r_i/d_i > 1$. Under $h_i^{(1)} = (1 - r_i/d_i)_+$, the penalty for a shortage is zero, so 8 vehicles are transferred out to the 4th station, which is not reasonable. Under $h_i^{(2)} = exp\{-r_i/d_i\}$, because the penalty is larger than zero, no vehicles are transferred out. For the 9th station, the prediction for demand is 9 and for return is 10, so that r_i/d_i is slightly larger than one. Under $h_i^{(1)} = (1 - r_i/d_i)_+$, no bicycles are transferred in. When $h_i^{(2)} = exp\{-r_i/d_i\}$, 7 bicycles from the 21st and 10th stations are transferred, with larger values of r_i/d_i (Table 5).

We find no bicycle transfer if the penalty cost for a rental station has an inverse linear relationship with the ratio of returns to rentals. Nevertheless, when the penalty cost is exponentially dependent on the negative ratio of returns to rentals, bicycle transfer occurs between rental stations with large ratios. This is because large ratios lead to high penalty costs, but bicycle transfer can reduce the penalty costs.

Table 4. Closed rental centers in each cluster

Cluster	Closed	Station ID									
1	10/20	3036	3181	3182	3252	3254	3326	3330	3333	3337	3395
2	1/40	329									
3	0/35	–									
4	0/29	–									
5	10/45	137	290	318	352	367	464	510	538	3234	3264
6	5/37	223	404	405	462	463					
7	5/34	3044	3053	3061	3063	3229					
8	2/31	3219	3250								
9	4/32	218	233	314	3246						
10	0/26	–									
11	2/25	3401	3432								
		3452	3453	3454							
13	2/30	3287	3291								
14	1/32	3385									
15	6/42	250	263	294	300	375	403				
16	6/36	3133	3138	3149	3153	3154	3257				
17	6/28	271	298	431	3222	3253	3455				
18	1/28	2005									
19	1/20	3237									
20	5/38	489	512	521	3017	3450					

Table 5. Demand and return forecasts in February 22-th, 2017 in Group 11

	1	2	3	4	5	6	7	8	9	10	11	12	13
Demand	19	26	24	24	16	14	11	19	9	12	35	19	20
Return	16	18	23	19	19	14	12	19	10	15	34	22	22
r_i/d_i	0.84	0.69	0.96	0.79	1.19	1.00	1.09	1.00	1.11	1.25	0.97	1.16	1.10
	14	15	16	17	18	19	20	21	22	23	24	25	
Demand	32	27	23	9	36	7	3	18	14	11	56	0	
Return	30	24	30	14	39	10	2	23	16	13	63	0	
r_i/d_i	0.94	0.89	1.30	1.56	1.08	1.43	0.67	1.28	1.14	1.18	1.13	0.00	

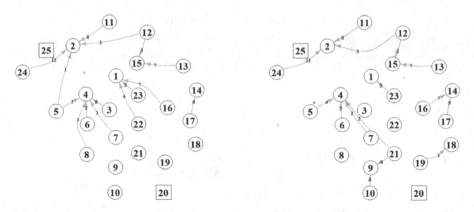

Fig. 6. Bicycle transfer plans in the setting of Group 11, i.e., $h_i^{(1)} = (1 - r_i/d_i)_+$(left) and $h_i^{(2)} = exp(-r_i/d_i)$(right)

5   Conclusions

This paper studied a bike-sharing transport network design problem in which both bicycle demand and return are uncertain. Bicycles are transferred among different stations to meet customer demand. We have considered two kinds of demand: regular and weather-related. Moreover, we used multi-order exponential smoothing methods to address seasonality and demand-return forecasting trends. We then applied machine learning techniques to handle weather-related demand and return patterns. To determine the optimal locations of rental stations, we formulated a stochastic mixed-integer programming problem and developed a Benders decomposition-based algorithm. Finally, we considered a real case, the NYC Citi Bike Program, to test our model and algorithm.

There are possible extensions. One is to introduce repair/maintenance into our model, where it takes some time to repair or maintain bicycles with a certain probability. The other is to improve the Benders decomposition-based algorithm that we proposed. Nevertheless, while data sets become large, it becomes time-consuming to find a good, feasible solution using this algorithm. Therefore, developing a more efficient algorithm for solving bike-sharing network design problems in big-data settings is necessary. Also, DBSCAN may be more appropriate than K-mean clustering as the number of clustering and shape of clustering are unknown. Another possible extensions is to compare our proposed model with existing work to give insights on the practical use of time series and random forest models. Finally, it is possible to extend current locations to other unused locations, thus developing a perhaps better bike-sharing network.

References

Citibike (2017). https://www.citibikenyc.com/system-data

Cruz, F., Subramanian, A., Bruck, B., Lori, M.: A heuristic algorithm for a single vehicle static bike sharing rebalancing problem. Comput. Oper. Res. **79**, 19–33 (2017)

Dell'Amico, M., Novellani, S., Stutzle, T.: A destroy and repair algorithm for the bike sharing Rebalancing Problem. Comput. Oper. Res. **71**, 149–162 (2016)

Forma, I., Raviv, T., Tzur, M.: A 3-step math heuristic for the static repositioning problem in bike-sharing systems. Transp. Res. Part B: Methodol. **71**, 230–247 (2015)

Gao, X., Lee, G.M.: Moment-based rental prediction for bicycle-sharing transportation systems using a hybrid genetic algorithm and machine learning. Comput. Industr. Eng. **128**, 60–69 (2019)

Ghosh, S., Adulyasak, Y., Jaillet, P.: Dynamic repositioning to reduce lost demand in bike sharing systems. J. Artif. Intell. Res. **58**, 387–430 (2017)

Kadri, A.A., Kacem, I., Labadi, K.: A branch-and-bound algorithm for solving the static rebalancing problem in bicycle-sharing systems. Comput. Industr. Eng. **95**, 41–52 (2017)

Larsen, J., Patterson, Z., El-Geneidy, A.: Build it. But where? The use of geographic information systems in identifying locations for new cycling infrastructure. Int. J. Sustain. Transp. **7**(4), 299–317 (2013)

Li, Y., Szeto, W., Long, J., Shui, C.: A multiple type bike repositioning problem. Transp. Res. Part B: Methodol. **90**, 263–278 (2016)

Lin, J., Yang, T.: Strategic design of public bicycle sharing systems with service level constraints. Transp. Res. Part E: Logist. Transp. Rev. **47**(2), 284–294 (2011)

Ho, S., Szeto, W.: A hybrid large neighborhood search for the static multi-vehicle bike-repositioning problem. Transp. Res. Part B: Methodol. **95**, 340–363 (2017)

Santoso, T., Shabbir, A., Goetschalckx, M., Shapiro, A.: A stochastic programming approach for supply chain network design under uncertainty. Eur. J. Oper. Res. **167**(1), 96–115 (2005)

Szeto, W., Shui, C.: Exact loading and unloading strategies for the static multi-vehicle bike repositioning problem. Transp. Res. Part B: Methodol. **109**, 176–211 (2018)

Vogel, P.: Service network design of bike sharing systems. In: Service Network Design of Bike Sharing Systems. Lecture Notes in Mobility, pp. 113–135. Springer, Cham (2016). https://doi.org/10.1007/978-3-319-27735-6_6

Yan, S., Lin, J.R., Chen, Y.C., Xie, F.R.: Rental bike location and allocation under stochastic demands. Comput. Industr. Eng. **107**, 1–11 (2017)

What are Smart Home Product Users Commenting on? A Case Study of Robotic Vacuums

Yixiu Yu[1]([⊠]), Qian Fu[2], Dong Zhang[3], and Qiannong Gu[1]

[1] Ball State University, Muncie IN, USA
{yyu4,qgu}@bsu.edu
[2] University of Birmingham, Birmingham, United Kingdom
q.fu@bham.ac.uk
[3] Dalian University of Technology University, Dalian, China
zhangdong@dlut.edu.cn

Abstract. Researchers are increasingly focusing on smart home products, yet a research gap remains in understanding user satisfaction and dissatisfaction with these products. Robotic vacuums, as a prevalent smart home product type, have become integral to daily life. In this study, we employed topic modeling to extract key factors influencing user satisfaction and dissatisfaction from online reviews of robotic vacuums on Amazon.com. We found that suction power, mapping, and mopping capability are the top factors contributing to user satisfaction, whereas complaints frequently centered around poor mapping, poor customer service, and connectivity issues. We note that mapping emerges as a double-edged sword that can either enhance satisfaction or generate dissatisfaction with robotic vacuums if performed poorly. Our research aims to bridge the existing knowledge gap, adding richness to the broader field of smart home literature. Also, our findings offer practical guidance for practitioners designing smart home products to enhance user satisfaction and minimize dissatisfaction.

Keyword: Smart home products Robotic vacuums User satisfaction User dissatisfaction Online reviews Topic modeling

1 Introduction

In recent years, there has been an increasing household penetration of smart home devices that are enabled by the Internet of Things (IoT). Smart home products have three critical components: physical components, "smart" components, and connectivity components [1]. These new products provide users with many benefits, such as efficiency and convenience but also generate some privacy and security concerns [2]. Therefore, users' feelings toward using smart home products are often quite mixed with both satisfaction and dissatisfaction. However, there is a lack of knowledge of the underlying influencing factors of smart home user satisfaction and dissatisfaction.

H. Han and E. Baker (Eds.): SDSC 2023, CCIS 2113, pp. 34–45, 2024.
https://doi.org/10.1007/978-3-031-61816-1_3

Online reviews have been regarded as a useful source for knowing users' actual experiences. In this study, we extracted influencing factors from users' online reviews on smart vacuums, a common overarching category. This research addresses the following research question: What are the key factors of robotic vacuum user satisfaction and dissatisfaction expressed in users' reviews?

1.1 Smart Home Research in IS

Researchers increasingly pay attention to smart home products. We have summarized extant IS literature on smart homes in Table 1. Overall, these studies emerged in recent years, from 2020 to 2023. The main research method is online survey. The extant research on smart homes mainly focuses on user intention and underlying drivers. Limited attention has been paid to users' actual experience and post-adoption usage. A research gap still exists in understanding smart home users' satisfaction and dissatisfaction and their influencing factors.

Table 1. A summary of extant smart home literature in the IS field

Literature	Journal	Context	Method	Major findings
[3]	CHB	Intelligent virtual assistants	Survey	The authors found that perceived authenticity and personalization have positive impacts on commitment as well as trust. They also found that user involvement and connection with the intelligent virtual assistants mediate the relationship between trust, commitment and reusage intentions
[4]	JCIS	General smart home devices	Survey	The authors found that performance expectancy, effort expectancy, social influence, hedonic motivation, price value, and trust positively influence individuals' intentions to use smart home devices. Perceived security risk has a negative impact on the intentions

(*continued*)

Table 1. (*continued*)

Literature	Journal	Context	Method	Major findings
[5]	ISJ	Smart home assistants (SHAs)	Online experiment and survey	The authors find that SHAs' anthropomorphic features can mitigate the intrusive effects of SHAs on users and their social relations at home
[6]	CHB	Voice-activated smart home devices	Survey	Users' opinion-seeking behavior has an influence on their perceptions of device utility for online retail activity
[7]	IJHCS	Smart home healthcare technology	Focus groups	The older users who have tried and used smart home monitoring technology would have fewer privacy concerns compared to non-use older users. They also found if smart home monitoring technologies allow for customized functionality and features, they would be more acceptable to those older users
[8]	C&S	General smart home devices	Literature review	The authors presented an overview of the vulnerabilities, risks, and countermeasures of smart homes
[9]	JBR	Voice assistants	Survey and sentiment analysis	Brand credibility moderates the relationship between VA features and the overall perceived value of VAs. Higher brand credibility mitigates users' perception of privacy risks

(*continued*)

Table 1. (*continued*)

Literature	Journal	Context	Method	Major findings
[10]	CHB	Smart speaker-based voice assistants	Longitudinal field deployment study	Older adults' perceived benefits of the voice assistants changed from initially enjoying simplicity and convenience of operation to not worrying about making mistakes and building digital companionship when they are familiar with the devices. The initial challenge is the unfamiliarity with the device, and the later challenge is coping with the functional errors
[11]	C&S	General smart home devices	Survey	Perceived security risk has an impact on users' intentions to use smart home devices, mediated by attitude and perceived control over the secure use of smart home devices
[12]	JAIS	Smart personal assistants (SPAs)	Literature review	The authors reviewed literature and web resources related to SPAs and developed a taxonomy. Using cluster analysis, the authors group SPAs and examine how different material properties of SPAs can influence affordances for value co-creation
[13]	CHB	General smart home devices	Experiment	The authors introduce two types of social connectedness that are related to the interaction between users and smart home devices: Inner Social Connectedness (ISC) and Outer Social Connectedness (OSC). They found that both ISC and OSC can increase users' perceived social support, which, in turn, positively influences companionship with smart home devices

(*continued*)

Table 1. (*continued*)

Literature	Journal	Context	Method	Major findings
[14]	IJIM	Smart thermostats	Content analysis and survey	The authors developed a novel factor, "techno-coolness," that has a positive effect on the smart thermostat adoption intention
[15]	IT&P	Smart locks	Survey	The perceived relative advantage of smart locks over conventional locks in providing safety and security is a significant factor of user adoption intention for both males and females. The perceived novel benefit is a significant factor of adoption intention for females but not for males
[16]	ISF	General smart home devices	Survey	The authors found that smart home users' post-disconfirmation dissonance can generate feelings of anger, guilt, and regret, related to dissonance reduction mechanisms, which in turn influence user satisfaction and well-being
[17]	CHB	Smart voice assistants	Survey	Smart voice assistant users are motivated by the products' utilitarian benefits, symbolic benefits, and social benefits. Perceived privacy risk is a moderator in dampening the use of smart voice assistants
[18]	IJIM	smart voice assistants	Survey	This paper examines the impacts of hedonic and utilitarian attitudes on smart voice assistant usage and word-of-mouth (WOM) recommendations. It also examines the influences of playfulness, escapism, anthropomorphism, visual appeal, and social presence on both attitudes

(*continued*)

Table 1. (*continued*)

Literature	Journal	Context	Method	Major findings
[19]	B&IT	digital personal assistants	Qualitative thematic analysis	This paper derived six main themes: Mundane tasks, the connected home, personification, family context, usability, and security
[20]	ISF	General smart home devices	Survey	Functional value and emotional value are associated with users' intention to stay in smart accommodations. Factors such as price, perceived usefulness, and perceived control have influences on the functional value and emotional value of smart accommodations
[21]	ISF	Smart home services	Survey	The authors found that user hedonic motivation moderates the impact of behavioral intention and user behavior. It is also found that the use of IoT smart home services has a positive influence on user well-being. Hedonic and eudaimonic motives' influences on user well-being is mediated by the use of smart home services
[22]	IJHCI	General smart home devices	Survey	The authors found that the knowledge of, perceived benefits, adoption of, and use of smart home products are related
[23]	IJHCI	Smart home voice assistants	Semi-structured in-depth interviews	Discontinuance and restricted acceptance can reduce users' worry about smart home voice assistants

(*continued*)

Table 1. (*continued*)

Literature	Journal	Context	Method	Major findings
[24]	IJHCS	Smart thermostats and networked lighting	Content analysis	Drawn evidence from literature reviews, the authors found the articles on smart homes promote the idea of enhanced quality of life, i.e., the vision of pleasance, and they revealed seven main qualities of pleasance: (1) aesthetic experience, (2) fun and cool, (3) customization and control, (4) convenience and simplicity, (5) peace of mind, (6) extension and expansion, and (7) effortless energy-saving
[2]	JCIS	Smart home devices	Survey	The authors found that performance expectancy and compatibility positively influence users' intentions to adopt smart home, and privacy risk, performance risk, and time risk negatively influence the adoption intention
[25]	JCIS	smart home assistants	Survey	The authors find that the use of different types of smart home assistants leads to different levels of gratification in different categories, which positively influence users' continuance intention. Age moderates the relationship between the use of smart home assistants and loneliness
[26]	I&M	Smart surveillance	Survey	This paper examines the trade-offs of contextual personalization and privacy concerns underlying smart surveillance on user behavioral intention

2   Data and Methods

In this section, we describe the data collection, data pre-processing and data, and the adopted text mining techniques. Figure 1 shows the text-mining framework in this study.

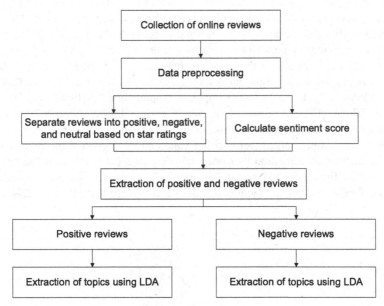

Fig. 1. Text-mining framework in this study

2.1   Data Collection and Pre-Processing

We chose smart vacuum/robot vacuum/robotic vacuum products as an example category of smart home products and downloaded their users' online reviews from Amazon.com using a web crawler in July 2021. There were 138 products, and the initial data set consisted of 116,677 reviews. This study considered only English reviews and omitted non-English reviews.

We used Python to pre-process and analyze the user review data. The Python VADER library has some advantages for sentiment analysis, including considering heuristics and retaining a review's original structure [27]. Therefore, in this study, we used the VADER library to detect sentiments of reviews. This research excluded neutral reviews and analyzed positive and negative reviews, which were categorized based on both star ratings and sentiment scores.

2.2 Topic Modeling

Following prior literature [27], we used the pyLDAvis package in Python to plot inter-topic distance maps (Fig. 2 and Fig. 3). Researchers can determine a good LDA model if it scatters relatively large and non-overlapping bubbles throughout the chart, not clustered in one quadrant [27].

We identified three topics from positive reviews (that is, suction power, mapping, and mopping capability) and three topics from negative reviews (that is, poor mapping, poor customer service, and connectivity issue). As suggested by previous researchers [28], the topic modeling is based on the group evaluation of the logical semantic relations of top words in each topic. Topic names were generated from the logical connection among the keywords (Table 2). For example, the topic name "suction power" captures such keywords as "Clean," "time," "dog_hair," "pick," "house," "carpet," "cleaning," "dust," "suction," "clean_everyday," "life_saver," "wish_soon," "amazed_pick," "hairy," and "clean_hair." Topic names were further validated by analyzing the representative reviews of each topic, a method that prior scholars used [28]. The topic names were finalized and confirmed until all researchers agreed. Topic names, i.e., the key attributes of user satisfaction and dissatisfaction, and the keywords are presented in Table 2.

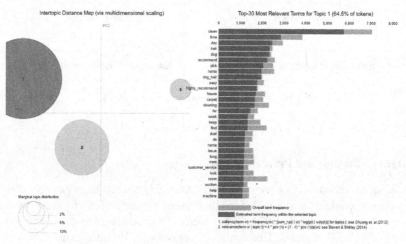

Fig. 2. Inter-topic distance map for positive reviews

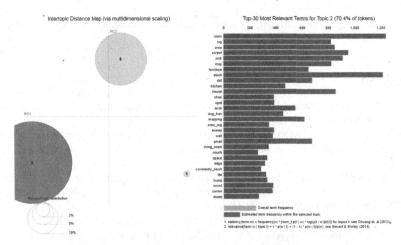

Fig. 3. Inter-topic distance map for negative reviews

Table 2. Topics and keywords extracted by the LDA model from robotic vacuum reviews

Type of reviews	Topic name	Keywords
Positive reviews	Suction power	Clean, time, dog_hair, pick, house, carpet, cleaning, dust, suction, clean_everyday, life_saver, wish_soon, amazed_pick, hairy, clean_hair
	Mapping	Keep_zone, map, room, area, find, know, clean_zone, map_create, remap, avoidance, add_room
	Mopping capability	Mop, pad, jet, water, clean, reusable_pad, disposable_pad, washable_pad, authentic_part, change_pad
Negative reviews	Poor mapping	Room, stuck, carpet, rug, pick, area, map, furniture, mapping
	Poor customer service	Customer service, replacement, warranty, service, defective, send_new, email, repair, send_replacement, replacement_part
	Connectivity issues	network, suit, stay_far, misdeed, financial. Nearly_year

3   Conclusion

LDA was employed to extract topics from positive and negative product reviews separately and identify influencing factors of robotic vacuum user satisfaction and dissatisfaction. This study can provide insights for practitioners regarding smart product design. It can also contribute to the smart home research and online review literature. The text mining techniques used in this study can be generalized and applied to other product reviews.

References

1. Porter, M.E., Heppelmann, J.E.: How smart, connected products are transforming competition. Harv. Bus. Rev. **92**, 64–88 (2014)
2. Wang, X., McGill, T.J., Klobas, J.E.: I want it anyway: consumer perceptions of smart home devices. J. Comput. Inf. Syst. **60**, 437–447 (2020). https://doi.org/10.1080/08874417.2018.1528486
3. Alimamy, S., Kuhail, M.A.: I will be with you Alexa! The impact of intelligent virtual assistant's authenticity and personalization on user reusage intentions. Comput. Hum. Behav. **143**, 107711 (2023). https://doi.org/10.1016/j.chb.2023.107711
4. Aldossari, M.Q., Sidorova, A.: Consumer acceptance of internet of things (IoT): smart home context. J. Comput. Inf. Syst. **60**, 507–517 (2020). https://doi.org/10.1080/08874417.2018.1543000
5. Benlian, A., Klumpe, J., Hinz, O.: Mitigating the intrusive effects of smart home assistants by using anthropomorphic design features: a multimethod investigation. Inf. Syst. J. **30**, 1010–1042 (2020). https://doi.org/10.1111/isj.12243
6. Canziani, B., MacSween, S.: Consumer acceptance of voice-activated smart home devices for product information seeking and online ordering. Comput. Hum. Behav. **119**, 106714 (2021). https://doi.org/10.1016/j.chb.2021.106714
7. Ghorayeb, A., Comber, R., Gooberman-Hill, R.: Older adults' perspectives of smart home technology: are we developing the technology that older people want? Int. J. Hum.-Comput. Stud. **147**, 102571 (2021). https://doi.org/10.1016/j.ijhcs.2020.102571
8. Hammi, B., Zeadally, S., Khatoun, R., Nebhen, J.: Survey on smart homes: vulnerabilities, risks, and countermeasures. Comput. Secur. **117**, 102677 (2022). https://doi.org/10.1016/j.cose.2022.102677
9. Jain, S., Basu, S., Dwivedi, Y.K., Kaur, S.: Interactive voice assistants – does brand credibility assuage privacy risks? J. Bus. Res. **139**, 701–717 (2022). https://doi.org/10.1016/j.jbusres.2021.10.007
10. Kim, S., Choudhury, A.: Exploring older adults' perception and use of smart speaker-based voice assistants: a longitudinal study. Comput. Hum. Behav. **124**, 106914 (2021). https://doi.org/10.1016/j.chb.2021.106914
11. Klobas, J.E., McGill, T., Wang, X.: How perceived security risk affects intention to use smart home devices: a reasoned action explanation. Comput. Secur. **87**, 101571 (2019). https://doi.org/10.1016/j.cose.2019.101571
12. Knote, R., Janson, A., Sollner, M., Leimeister, J.M.: Value co-creation in smart services: a functional affordances perspective on smart personal assistants. J. Assoc. Inf. Syst. **22**(2), 418–458 (2021). https://doi.org/10.17705/1jais.00667
13. Lee, B., Kwon, O., Lee, I., Kim, J.: Companionship with smart home devices: the impact of social connectedness and interaction types on perceived social support and companionship in smart homes. Comput. Hum. Behav. **75**, 922–934 (2017). https://doi.org/10.1016/j.chb.2017.06.031
14. Mamonov, S., Koufaris, M.: Fulfillment of higher-order psychological needs through technology: the case of smart thermostats. Int. J. Inf. Manag. **52**, 102091 (2020). https://doi.org/10.1016/j.ijinfomgt.2020.102091
15. Mamonov, S., Benbunan-Fich, R.: Unlocking the smart home: exploring key factors affecting the smart lock adoption intention. Inf. Technol. Amp People. (2020). https://doi.org/10.1108/ITP-07-2019-0357

16. Marikyan, D., Papagiannidis, S., Alamanos, E.: Cognitive dissonance in technology adoption: a study of smart home users. Inf. Syst. Front. (2020). https://doi.org/10.1007/s10796-020-100 42-3

17. McLean, G., Osei-Frimpong, K.: Hey Alexa … examine the variables influencing the use of artificial intelligent in-home voice assistants. Comput. Hum. Behav. **99**, 28–37 (2019). https://doi.org/10.1016/j.chb.2019.05.009

18. Mishra, A., Shukla, A., Sharma, S.K.: Psychological determinants of users' adoption and word-of-mouth recommendations of smart voice assistants. Int. J. Inf. Manag. **67**, 102413 (2022). https://doi.org/10.1016/j.ijinfomgt.2021.102413

19. Paay, J., Kjeldskov, J., Hansen, K.M., Jørgensen, T., Overgaard, K.L.: Digital ethnography of home use of digital personal assistants. Behav. Inf. Technol. **41**, 740–758 (2022). https://doi.org/10.1080/0144929X.2020.1834620

20. Papagiannidis, S., Davlembayeva, D.: Bringing smart home technology to peer-to-peer accommodation: exploring the drivers of intention to stay in smart accommodation. Inf. Syst. Front. (2021). https://doi.org/10.1007/s10796-021-10227-4

21. Sequeiros, H., Oliveira, T., Thomas, M.A.: The impact of IoT smart home services on psychological well-being. Inf. Syst. Front. (2021). https://doi.org/10.1007/s10796-021-101 18-8

22. Shank, D.B., Wright, D., Lulham, R., Thurgood, C.: Knowledge, perceived benefits, adoption, and use of smart home products. Int. J. Human-Computer Interact. **37**, 922–937 (2021). https://doi.org/10.1080/10447318.2020.1857135

23. Shank, D.B., Wright, D., Nasrin, S., White, M.: Discontinuance and restricted acceptance to reduce worry after unwanted incidents with smart home technology. Int. J. Human–Comput. Interact. **39**(14), 2771–2784 (2022). https://doi.org/10.1080/10447318.2022.2085406

24. Strengers, Y., Hazas, M., Nicholls, L., Kjeldskov, J., Skov, M.B.: Pursuing pleasance: interrogating energy-intensive visions for the smart home. Int. J. Hum.-Comput. Stud. **136**, 102379 (2020). https://doi.org/10.1016/j.ijhcs.2019.102379

25. Xie, Y., Zhao, S., Zhou, P., Liang, C.: Understanding continued use intention of AI assistants. J. Comput. Inf. Syst. **63**(6), 1424–1437 (2023). https://doi.org/10.1080/08874417.2023.216 7134

26. Zhang, F., Pan, Z., Lu, Y.: AIoT-enabled smart surveillance for personal data digitalization: contextual personalization-privacy paradox in smart home. Inf. Manage. **60**, 103736 (2023). https://doi.org/10.1016/j.im.2022.103736

27. Lee, C.K.H.: How guest-host interactions affect consumer experiences in the sharing economy: new evidence from a configurational analysis based on consumer reviews. Decis. Support. Syst. **152**, 113634 (2022). https://doi.org/10.1016/j.dss.2021.113634

28. Ding, K., Choo, W.C., Ng, K.Y., Ng, S.I., Song, P.: Exploring sources of satisfaction and dissatisfaction in Airbnb accommodation using unsupervised and supervised topic modeling. Front. Psychol. **12**, 659481 (2021). https://doi.org/10.3389/fpsyg.2021.659481

Detecting Political Polarization Using Social Media Data

Erdogan Dogdu[1]([✉]), Roya Choupani[1], and Selim Sürücü[2]

[1] Department of Computer Science, Angelo State University, San Angelo, TX, USA
{edogdu,rchoupani}@angelo.edu
[2] Department of Computer Engineering, Çankırı Karatekin University,
Çankırı, Turkey
selimsurucu@karatekin.edu.tr

Abstract. Political polarization is a world wide problem observed everywhere. It is on the increase due to increasing digital communication and use of social media. Election times are especially important, increasing political polarization tremendously, causing social and political problems in the society. In this study, we take the case of last Turkish local elections, which was highly contentious, leading to the renewal of elections in the most-populated major metropolitan city of Istanbul with a population of 15 million people. We used Twitter data to measure the political polarization, collected over 90 million tweets and retweets, introduced the inter retweet ratio metric to quantitatively measure the political polarization. Our results show that political polarization increases tremendously by double digit decreases in inter community interactions towards the election days, and comes back to normal after the elections are over.

Keywords: Political polarization · community detection · social network analysis · big data

1 Introduction

Political polarization refers to the growing ideological or political distance or alienation between individuals or communities who hold different political views. It happens when people increasingly identify with a particular political party or ideology and become more entrenched in their beliefs, leading to greater hostility and less willingness to compromise with those who hold opposing views.

Polarization can be fueled by a variety of factors, including economic inequality, media fragmentation, demographic changes, or social media algorithms, as observed in the case of Cambridge Analytica[1], that create echo chambers. As people become more insulated from alternative viewpoints, they may perceive those with differing beliefs as hostile and even further distance themselves.

The results of political polarization can be damaging for society. It can make it difficult for individuals to engage in productive dialogue and compromise,

---

[1] https://www.theguardian.com/technology/2018/mar/17/facebook-cambridge-analytica-kogan-data-algorithm.

H. Han and E. Baker (Eds.): SDSC 2023, CCIS 2113, pp. 46–59, 2024.
https://doi.org/10.1007/978-3-031-61816-1_4

leading to gridlock in government and a lack of progress on important issues. Polarization can also lead to increased hostility and even violence between groups with different political views. It can erode trust in institutions and reduce social cohesion, making it harder for people to work together to solve problems.

In summary, political polarization is a phenomenon that occurs when people become more ideologically entrenched and hostile towards those with different political views. It can result in gridlock in government, increased hostility between groups, and a lack of progress on important issues.

Political polarization can be measured using social networks by analyzing the patterns of communication and interactions between users with different political views. One of the main social platforms used in such studies in Twitter. One way to measure political polarization on Twitter is by examining the level of agreement or disagreement between users on political topics. This can be done by analyzing the content of tweets using natural language processing techniques to identify keywords and sentiment. Another way to measure political polarization on Twitter is by analyzing the network structure of users and their connections. This can be done using network analysis tools to identify clusters of users with similar political views and to measure the degree of separation between these clusters. This second approach measures the interaction among users and groups, which is important in measuring the polarization.

A commonly used method to measure political polarization on Twitter is the "echo chamber" approach [2], which looks at the extent to which users are exposed to diverse viewpoints or are instead primarily interacting with like-minded individuals. This can be done by analyzing the retweeting and liking patterns of users, as well as the content of the tweets they engage with.

In this study we take the network structure approach [7] to study political polarization for the case of 2019 Turkish local elections. We use Twitter data that is collected in a 6-months period covering the 2019 elections. Therefore, this is long-term study. We collected about 90 million tweets and retweets. Therefore, this is a large-scale study. We identify political groups using community detection algorithms and we mainly use a graph-oriented approach by representing users as nodes and retweet interactions among users as weighted directed edges. Our main claim to show political polarization quantitatively is to show that interactions, or retweeting, among Twitter communities are decreasing during elections, therefore showing the increase of political polarization.

This paper is organized as follows: After the introduction, Sect. 2 presents the related work in this area. Section 3 presents the graph-based methods used in community detection. In Sect. 4, we present the polarization detection method we propose and the test results. Finally, we conclude in Sect. 5 and point to future work.

2 Related Work

Political polarization is a topic of interest that has been studied among political scientists for a long time. In recent years political polarization is also studied by data scientists due to the growing effect of internet and the digital platforms like social media. Now it is much more easier than ever to collect large volumes of big data from online social networks and study political polarization from multiple aspects and extract insights about the topic in general or certain problems in detail.

Twitter is one of the widely used social media platforms used to study political polarization as well as many social science experiments, especially using big data and data science methods. Here we review some of the relevant work.

Garimella et al. [9] collected data from Twitter accounts that are politically-oriented. They analyzed the change in polarization over years by collecting data from a large set of 679,000 user accounts between two elections in the US. It has been shown that political polarization has increased by 10% to 20% from 2009 to 2016 period.

In another study, Waller et al. [26] studied Reddit comments, 5.1B comments made in 10K communities over 14 years. They developed a neural embedding methodology to quantitatively measure the positioning of online communities along social dimensions by leveraging large-scale patterns of aggregate behavior. According to the results, Reddit underwent a significant polarization period after 2016 elections and remained polarized afterwards.

Darwish [8] quantified polarization over retweet and hashtag data using semi-supervised and moderated classification among more than 128,000 Twitter users during Brett Kavanaugh's nomination to the U.S. Supreme Court. In this study, polarization is analyzed by modifying existing polarization techniques.

In another study, Rashed et al. [5] aimed to measure the political polarization in the summer of 2013 in terms of change of opinion on 2 axes (Islamic/secular, pro-military/military intervention). Little evidence was found that humans switched sides between these 2 axes. However, it has been revealed that the thoughts belonging to these 2 axes have become more active in different periods.

Rashed et al. [21] analyzed the 2018 presidential election in Turkey, with over 213 million tweets supporting or opposing Recep Tayyip Erdoğan. It was stated that user stances were 90% certain. They targeted many clusters, including political figures, different groups and parties. Using a CNN-based multilingual universal sentence coder, the analysis calculated the polarization between the subjects. However, their study is limited to 168K user in comparison to the originally collected 653K user. In our case we worked with 384K unique users. They also collected tweets going backward starting in Dec 2018 towards June 2018 when the elections we held. In our case, we collected tweets starting before the elections and stopping after a few months the elections were over. So, our coverage is more distributed and reflecting a wider population.

Lada et al. [1] on the other hand studied the US Presidential elections in 2004 and focused on measuring the degree of interaction between liberal and conservative bloggers, and uncovering differences in the structure of the two groups. Strong bloggers and the individuals were identified in this study. Their effects were measured as in the right or left. In this study on blog pages, it is shown that the rates of political influence can be determined.

Morales et al. [16] measure the effect of tweets, sent by President Hugo Chavez in Venezuelan elections, on Twitter users and show how polarization is realized. They detect different degrees of political polarization by using different network structures, and the reactions of users to the tweets. The method they developed is called visual intensity function.

3   Community Detection in Social Networks

A first step in detecting political polarization in a social network is to detect and identify communities within that social network. Communities in a social network represent are those people who are in close communication or in a somewhat higher bandwidth of messaging among themselves. Our aim is first to identify those distinct communities and then measure the bandwidth among those communities through time over a period of specific political process such as elections.

Community detection is a widely studied topic with many methods developed over time. Here we summarize some of the related methods.

3.1   Graph Representation of Social Networks

Social networks can best be represented using "graphs". Graphs are simply connected nodes representing a network. In this case, users in a social network are represented using connected nodes in a graph, and the social interaction between users users can be represented using edges or links between users [23]. Graphs are also used to represent complex structures in many areas such as communication networks, technological networks, biological networks, chemical interaction networks, and more.

Connections or relationships between nodes in a graph can be directional, originating from one node and going into another. These are called directed graphs and they represent the directional interaction between nodes. In the case of a social network, those edges or connections in a graph represent a user messaging or following or retweeting another user.

When connections are not directed, then edges in a graph represent some kind of a relationship in the social network, but it is not from one user to the other, but mutual. For example, in LinkedIn users form connections (mutual and non-directional), but they follow one another (directional).

Edges in a graph representing a social nework can be weighted, meaning there could be quantitative values associated with edges. Those numbers or weights can represent different things such as how many times a user retweeted another user.

3.2   Graph Measures and Metrics

There are many community detection algorithms developed over networks or graphs. These algorithms use various metrics when detecting communities or groups. Here, we first review some of these metrics.

Conductance: This is a metric for calculating how well-knitted a graph is. This metric calculates the density of paths from one community or group to another. If the result of this metric produces a value close to 0, it is considered a successful result, completely distinct communities [24]. However, this is usually not the case. The conductance of a community s is the ratio of edges going out from the community by the total number of edges connecting to the members of the community:

$$f(s) = \frac{oe_s}{e_s} \tag{1}$$

oe_s : The number of outgoing edges from s
e_s : The total number edges from community members

Erdos Renyi Modularity: Erdos Renyi Modularity, also known as Newman-Girvan Modularity, is based on the assumption that the vertices of randomly generated graphs are associated with a probability p [15]. It is a metric based on calculating the density of communities in a graph with probability values. It produces a result between 0 and 1.

$$Q(S) = \frac{1}{m} * \sum_{s \in S} \left[m_s - \frac{m * n_s * (n_s - 1)}{n * (n - 1)} \right] \tag{2}$$

where:

m : The number of edges in graph S
m_s : The number of edges in a community s
n_s : The number nodes in community s
n : The number of nodes in graph S

Modularity Density: This is a metric used to calculate the density of the number of nodes in each community rather than the total number of edges in the communities [6].

3.3   Community Detection Algorithms

Community detection on graphs is about finding strongly connected subsystems or groups of nodes in a graph. These subsystems are also called clusters or communities [13]. Community detection is used in many different fields such as social media [3,19], communications networks [4], biological systems and healthcare [28], economics [12], academic networks [18], e-commerce [22].

Many methods are developed to detect communities. Communities detected on a graph can be of two types: disjoint communities, where nodes are a member of only one community, and overlapping communities, where nodes can be members of multiple communities. Some of the well-known community detection algorithms are as follows:

Louvain Algorithm is a community detection algorithm proposed by Brondel et al. [4]. This algorithm, which aims to detect communities in large networks quickly, uses the modularity metric. The modularity metric, which appears as unitary in some sources, expresses the power of the components of the system to be taken apart and then reassembled. This metric produces values between −1 and +1. If a graph that has a modularity metric of −1 is considered the weakest, and if the metric is +1, then it is considered the strongest. Modularity metric values are calculated separately for each detected community. Then, it is tried to obtain the maximum value of the modularity metric calculated in each new community. The algorithm stops working if the calculated modularity metric does not change or reaches the maximum value. Louvain algorithm is used for weighted and unweighted graphs, and for directed and undirected graphs as well.

Label Propagation Algorithm (LPA), proposed by Raghavan et al. [20], is a semi-supervised machine learning algorithm that aims to assign tags to untagged nodes in a network structure. First, all nodes are initially assigned a unique tag. Then, the closeness values between the nodes are calculated. Closeness is a measure of centrality in a graph. When calculating this metric, it is calculated as the inverse of the sum of the lengths of the shortest paths between the node and all other nodes in the graph [11]. Then, nodes with the same or close proximity values are assigned the same label. The steps are repeated until the labels of the nodes remain constant (unchanged), and then the algorithm stops [27].

Girvan-Newman algorithm [10] is used to detect communities by using "edge betweenness" metric in a network structure [14]. The edge betweenness metric is based on the shortest paths between nodes in a graph, similar to the closeness metric. The most important difference of this metric is that it uses the number of shortest paths passing through edges. Girvan-Newman algorithm uses the metric to detect communities in an iteration by removing the edges with high edge betweenness values and recalculating the metric in each iteration.

The eigenvector algorithm [17] calculates eigenvectors using the modularity metric. Sub-communities are determined by looking at the densities between nodes. The value of the modularity metric is aimed to be maximum.

Leiden algorithm is an improved version of the Louvain algorithm proposed by Traag et al. [25]. The Leiden algorithm produces a more reliable solution in comparison to the Louvain algorithm. Nevertheless, it is more complex according to the Louvain algorithm, in terms of structure. This algorithm calculates the modularity value in sub-communities; it works with the logic of combining these sub-communities in order to maximize the modularity value. Larger ensembles are obtained using the algorithm. The algorithm works for all graph types, which is a big plus.

3.4 Dataset

We collected tweets during 2019 Turkish local elections[2]. We used Twitter API[3] for data collection from Twitter. Data is collected during a period of 157 days

---

[2] https://en.wikipedia.org/wiki/2019_Turkish_local_elections.

[3] https://developer.twitter.com/en/docs/twitter-api.

between 14 February 2019, and 20 July 2019. We collected about 90 million tweets, of which approximately 22 million (24%) are tweets and 68 million (76%) are retweets.

Twitter API allows us to sample tweet stream using pre-specified keywords. In our data collection process, we used the names of political candidates, head of political parties, the names of political parties, and a few election related keywords.

The 2019 Turkish local elections were highly contested, conducted over 81 provinces of Turkey on 31 March 2019. In the most populated major metropolitan city of Istanbul, with a population of over 15 million people, the top two mayoral candidates got votes very close to each other, the opposition party's candidate received only 14,000 more votes than the ruling party's candidate. The Election Board decided to renew the elections in Istanbul after the ruling party contested the result. Instanbul election is renewed on 23 June 2019. The opposition candidate won the renewed election with a landslide, 9% difference.

Accordingly, the election was very much contested. Social media, particularly Twitter, was very active, and it was apparent that people were very much polarized due to the heated discussions on the media, printed press and social media platforms. Here, we try to show the level of polarity with hard data from Twitter.

3.5 Evaluation

We first evaluate the community detection algorithms reviewed in Sect. 3. In our study, the graph representation we use are weighted directed graphs. The nodes represent Twitter users and edges in the graph represent retweet counts (weights) between users, from one user to the other, therefore the edges are directed. Not all community detection algorithms we reviewed support weighted and directed graphs. See Table 1 for a summary of the algorithms' support of different graph types. As seen, only three algorithms, namely LPA, Louvain, and Leiden, support weighted and directed graphs. Therefore, only these three algorithms can be used effectively for community detection. However, other algorithms can also be tested by transforming a directed weighted graph to unweighted and undirected graph.

Table 1. Community detection algorithms and their support of different graph types

Algorithms	Weighted Graph	Directed Graph
Girman-Newman	–	✓
LPA	✓	✓
Eigenvector	–	✓
Leiden	✓	✓
Louvain	✓	✓

We measured 3 metrics, reviewed above, on all 5 community detection algorithms. The results are presented in Table 2. Since the data we collected is very big to test, and the algorithms are not very scalable, we limited the tests to only tweets and retweets sent in the month of March 2019, in which the local elections took place. A total of 3.5 million tweets and retweets from 98 thousand users are used in the test.

Table 2. Performance Results of Community Detection Algorithms

Algorithms	Erdos Renyi Modularity	Conductance	Modularity Density
LPA	0.7829	0.1619	4190.1575
Leiden	0.8127	0.0061	3530.6777
Louvain	0.7887	0.0123	3498.3589
Girman-Newman	0.7111	0.0949	4395.4571
Eigenvector	0.3031	0.1120	3448.5386

According to the results in Table 2:

- There is no ideal value (best or worst) for the modularity density metric. The acceptable value for this metric is the maximum value. In this case, the algorithms' performance are ranked from best to worst as Girwan-Newman, LPA, Leiden, Louvain and Eigenvector.
- The conductance metric produces values between 0 and 1. 0 is considered the best value for this metric. All results are close to 0 and therefore all algorithms are usable.
- When we check the best value of the Erdos Renyi modularity metric, we see that it is 1. When we examine the results of the algorithms according to this value, all algorithms, except the eigenvector algorithm, produced acceptable results.

We also evaluated the number of communities found and the largest community's member count for each algorithm (Table 3). In most cases, the community detection algorithms put many users (nodes) in one large community due to big data and many transitive connections between users. That is something we want to avoid since we want to find as many interconnected communities as possible, so that we can detect polarization levels better.

Using the same smaller dataset, the number of communities found by the algorithms and the size of the largest community are listed in Table 3. A high number of communities mean that the number of members per group are lower, and therefore there will be too many interactions between communities. On the other hand, a lower number of communities indicates that there will be large groups in terms of the number of members, and in this case, there will be limited interaction between communities or groups. These two situations are problems that should be avoided as they may affect the analysis results in our study.

Leiden and Louvain algorithms give similar results. On the other hand, the eigenvector algorithm produces the opposite of the Leiden and Louvain algorithms, with a very large community with 82K members in comparison to 20K members in the case of Ledien and Louvain.

We decided to use LPA in this case. The results obtained are close to ideal case for all metrics (Erdos Renyi Modularity, Conductance, Modularity Density). It can also work on weighted graphs and directed graphs. In addition, based on the number of members of the groups with the LPA algorithm, neither a small number of members nor a large group was obtained compared to the other algorithms.

Table 3. The results of community detection algorithms

Algorithm	#Communities	Size of the largest community
LPA	11,242	28,919
Leiden	6,601	21,423
Louvain	6,637	22,711
Girman-Newman	11,433	21,506
Eigenvector	6,263	82,434

4   Political Polarization Detection in Social Networks

Political polarization is a common phenomena, recognized all over the world. Political polarization is broadly defined as "the divergence of political attitudes away from the center, towards ideological extremes"[4]. It is clear and obvious that increased digital communication over the internet, mainly via social networks, contribute to the increase of political polarization. To show the political polarization over a period of time during elections, we collected 90 million tweets and retweets during the 2019 Turkish local elections, then identified communities of users based on their interactions (retweets), and now we will show how these communities (groups) are interacting during the election period.

4.1   Political Polarization Metric

Our claim here is that if communities are interacting less during certain periods, such as election times, it is proof that there is political polarization, manifested by the decrease of inter community communication.

To be able measure the interaction among communities or groups in a social network, in this case Twitter, we introduce "inter-retweet ratio" metric, which is defined as the ratio of retweets among communities against the total number of retweets among all social network users.

---

[4] https://en.wikipedia.org/wiki/Political_polarization.

$$inter\_retweet\_ratio = \frac{\#inter\_retweets}{\#inter\_retweets + \#intra\_retweets} \qquad (3)$$

If the inter-retweet ratio decreases over a period of time, it is an indication that the communication among communities is decreasing, hence the political polarization is increasing.

4.2 Evaluation

The overall process is depicted in Fig. 1. It consists of three main steps: (1) data collection, (2) data preprocessing and community detection, and (3) political polarization detection.

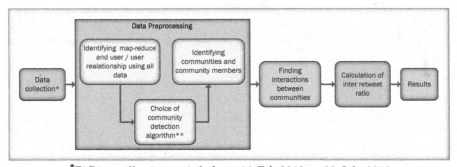

$^{*}=$ *Data collection period: from 14 Feb 2019 to 20 July 2019*
$^{**}=$ *The community detection algorithm was decided using March 2019 data.*

Fig. 1. Politcal Polarization Detection Process

Due to the size of data we collected, we resorted to the big data processing tools. We used Apache Spark[5] to process the tweets and retweets. We executed the processes on a a multi-core Linux server with 40 cores and 190GB RAM, employing 20 Spark workers. We also Python networkx[6] library for community detection implementation and pyspark[7] version of Spark.

Since we have been collecting data for a long period of time (6 months), there were many communities being detected. For this reason, we have added some criteria when identifying groups or communities. Since large communities mainly interact among themselves, not with other communities, we put an upper limit on the number of users in a community to be considered in the experiment. For the same reason, we also added a lower limit on the size of the communities to be considered in the experiment. The upper limit is 10,000 users and the lower limit is 1,000 users. With these limitations, the number of communities detected and the number of overall users obtained are listed in Table 4.

---

[5] https://spark.apache.org/.
[6] https://networkx.org/.
[7] https://spark.apache.org/docs/latest/api/python/.

Table 4. Number of communities and users detected

#Groups	4,767
#Unique users	384,272

In order to clearly see the effect of political polarization among Twitter communities or groups, we measured the inter-retweet ratio daily from 14 February 2019 to 20 July 2019. Figure 2) shows the results. Due to variations and irregularities in data (it is sampled), we also draw a 10-day moving average for the same period (orange line) to see the trend. Here are our observations:

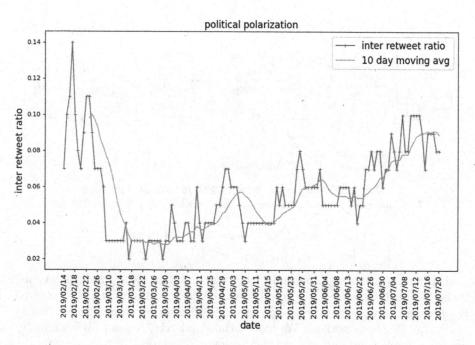

Fig. 2. Daily inter retweet ratios among Twitter political groups

- During the election period, the inter retweet ratio drops from the starting double digit rates (>.10) to almost .3.
- After the elections are over, the rate goes back to normal, again almost double digits.
- The lowest inter retweet ratio is observed on the election day week (31 Mar 2019).
- There is also a sharp decrease in inter retweet ratio on the election renew day week (20 June 2019).

These results show that political polarization is a reality. Users in different communities, possibly belonging to different political views, interact less (retweet each other less) during election period, specifically during and around election days.

5 Conclusions and Future Work

In this study, we have conducted a large scale and long-term experiment on political polarization by taking the case of 2019 Turkish local elections. We collected over 90 million tweets and retweets over a period of 6 months in which local elections and a renewed election in the most populated metropolitan city Istanbul took place. We identified Twitter political groups by using retweet counts among users. Then, using an inter retweet ratio metric, we proposed for measuring the political polarization, we observed daily changes of the metric over the same period. Results clearly show that political polarization increases tremendously among Twitter political communities by the decrease of interaction among these communities. We showed that there double digit drops in inter retweet ratios among political groups during election weeks. Luckily, the numbers also show that interaction among groups increases as the election days are over and things go back to normal, so there is hope.

We plan to continue this research by studying the upcoming US presidential elections in 2024 with a larger scale study covering one whole year including time before and after the elections, noting the important political events such as primaries, election days, inauguration, and observing how these events affect the political polarization and whether political polarization decreases after the elections are over.

References

1. Adamic, L.A., Glance, N.: The political blogosphere and the 2004 U.S. election: divided they blog. In: Proceedings of the 3rd International Workshop on Link Discovery, LinkKDD '05, pp. 36–43. Association for Computing Machinery, New York (2005). https://doi.org/10.1145/1134271.1134277
2. Barberá, P.: Social media, echo chambers, and political polarization. Social Media Democ. State Field Prospects Reform **34** (2020)
3. Bedi, P., Sharma, C.: Community detection in social networks. WIREs Data Min. Knowl. Disc. **6**(3), 115–135 (2016). https://doi.org/10.1002/widm.1178
4. Blondel, V.D., Guillaume, J.L., Lambiotte, R., Lefebvre, E.: Fast unfolding of communities in large networks. J. Stat. Mech. Theory Exp. **2008**(10), P10008 (2008). https://doi.org/10.1088/1742-5468/2008/10/p10008
5. Borge-Holthoefer, J., Magdy, W., Darwish, K., Weber, I.: Content and network dynamics behind Egyptian political polarization on twitter, CSCW 2015, pp. 700–711. Association for Computing Machinery, New York (2015). https://doi.org/10.1145/2675133.2675163
6. Chen, T., Singh, P., Bassler, K.E.: Network community detection using modularity density measures. J. Stat. Mech: Theory Exp. **2018**(5), 053406 (2018)

7. Conover, M., Ratkiewicz, J., Francisco, M., Gonçalves, B., Menczer, F., Flammini, A.: Political polarization on twitter. In: Proceedings of the International AAAI Conference on Web and Social Media, vol. 5, pp. 89–96 (2011)
8. Darwish, K.: Quantifying polarization on twitter: the kavanaugh nomination (2020). https://doi.org/10.48550/ARXIV.2001.02125
9. Garimella, K., Weber, I.: A long-term analysis of polarization on twitter (2017). https://doi.org/10.48550/ARXIV.1703.02769
10. Girvan, M., Newman, M.E.J.: Community structure in social and biological networks. Proc. Natl. Acad. Sci. **99**(12), 7821–7826 (2002). https://doi.org/10.1073/pnas.122653799
11. Golbeck, J.: Chapter 3 - network structure and measures. In: Golbeck, J. (ed.) Analyzing the Social Web, pp. 25–44. Morgan Kaufmann, Boston (2013). https://doi.org/10.1016/B978-0-12-405531-5.00003-1
12. Gui, X., Li, L., Cao, J., Li, L.: Dynamic communities in stock market. Abstr. Appl. Anal. **2014**, 723482 (2014). https://doi.org/10.1155/2014/723482
13. Javed, M.A., Younis, M.S., Latif, S., Qadir, J., Baig, A.: Community detection in networks: a multidisciplinary review. J. Netw. Comput. Appl. **108**, 87–111 (2018). https://doi.org/10.1016/j.jnca.2018.02.011
14. Lu, L., Zhang, M.: Edge Betweenness Centrality, pp. 647–648. Springer, New York (2013). https://doi.org/10.1007/978-1-4419-9863-7_874
15. McDiarmid, C., Skerman, F.: Modularity of erdős-rényi random graphs. Rand. Struct. Algor. **57**(1), 211–243 (2020). https://doi.org/10.1002/rsa.20910
16. Morales, A.J., Borondo, J., Losada, J.C., Benito, R.M.: Measuring political polarization: twitter shows the two sides of venezuela. Chaos Interdisc. J. Nonlinear Sci. **25**(3), 033114 (2015). https://doi.org/10.1063/1.4913758
17. Newman, M.: Finding community structure in networks using the eigenvectors of matrices. Phys. Rev. E, Stat. Nonlinear Soft Matter Phys. **74**, 036104 (2006). https://doi.org/10.1103/PhysRevE.74.036104
18. Oyelade, J., Oladipupo, O., Obagbuwa, I.: Application of k means clustering algorithm for prediction of students academic performance. Int. J. Comput. Sci. Inf. Secur. **7** (2010).https://doi.org/10.48550/arXiv.1002.2425
19. Ozer, M., Kim, N., Davulcu, H.: Community detection in political twitter networks using nonnegative matrix factorization methods. In: 2016 IEEE/ACM International Conference on Advances in Social Networks Analysis and Mining (ASONAM), pp. 81–88 (2016). https://doi.org/10.1109/ASONAM.2016.7752217
20. Raghavan, N., Albert, R., Kumara, S.: Near linear time algorithm to detect community structures in large-scale networks. Phys. Rev. E Stat. Nonlinear Soft Matter Phys. **76**, 036106 (2007). https://doi.org/10.1103/PhysRevE.76.036106
21. Rashed, A., Kutlu, M., Darwish, K., Elsayed, T., Bayrak, C.: Embeddings-based clustering for target specific stances: the case of a polarized turkey (2020). https://doi.org/10.48550/ARXIV.2005.09649
22. Ríos, S.A., Videla-Cavieres, I.F.: Generating groups of products using graph mining techniques. Procedia Comput. Sci. **35**, 730–738 (2014). https://doi.org/10.1016/j.procs.2014.08.155
23. Schaeffer, S.E.: Graph clustering. Comput. Sci. Rev. **1**(1), 27–64 (2007). https://doi.org/10.1016/j.cosrev.2007.05.001
24. Shi, J., Malik, J.: Normalized cuts and image segmentation. IEEE Trans. Pattern Anal. Mach. Intell. **22**(8), 888–905 (2000). https://doi.org/10.1109/34.868688
25. Traag, V.A., Waltman, L., van Eck, N.J.: From louvain to leiden: guaranteeing well-connected communities. Sci. Rep. **9**(1), 5233 (2019)

26. Waller, I., Anderson, A.: Quantifying social organization and political polarization in online platforms. Nature **600**(7888), 264–268 (2021). https://doi.org/10.1038/s41586-021-04167-x
27. Xing, Y., Meng, F., Zhou, Y., Zhu, M., Shi, M., Sun, G.: A node influence based label propagation algorithm for community detection in networks. Sci. World J. **2014**, 627581 (2014). https://doi.org/10.1155/2014/627581
28. Yanrui, D., Zhen, Z., Wenchao, W., Yujie, C.: Identifying the communities in the metabolic network using 'component' definition and girvan-newman algorithm. In: 2015 14th International Symposium on Distributed Computing and Applications for Business Engineering and Science (DCABES), pp. 42–45 (2015). https://doi.org/10.1109/DCABES.2015.18

Comparing Cost and Performance of Microservices and Serverless in AWS: EC2 vs Lambda

Christopher Allen[3], Xiaozhou Li[3,4], Amr S. Abdelfattah[1], Tomas Cerny[2(✉)], and Davide Taibi[3,4]

[1] Baylor University, Waco 76706, USA
`amr_elsayed1@baylor.edu`
[2] SIE, University of Arizona, Tucson, USA
`tcerny@arizona.edu`
[3] Tampere University, Tampere, Finland
`{christopher.allen,xiaozhou.li}@tuni.fi`
[4] Oulu University, Oulu, Finland
`{xiaozhou.li,davide.taibi}@oulu.fi`

Abstract. Serverless functions introduce a new way of running cloud-native systems suggesting a reduction in operational costs and delegating scalability to the cloud providers. However, while several companies are adopting serverless functions, it is still unclear if they have a competitive advantage compared to services deployed on docker containers. To compare the performance and the costs of serverless functions and microservices, we introduce a case study comparing microservice-based and serverless-based applications. We compare their performance and cost when each of its services is deployed in a docker container running on AWS EC2 or as a serverless function in AWS Lambda. Our study shows that the serverless functions performed better over time compared to the microservice version after initially having a slower response time. The microservice performance got worse over time, implying that the serverless functions are better suited for larger volumes of internet traffic. Results also indicated that the serverless functions are cheaper to operate than the microservices when the number of monthly requests is limited.

Keywords: Serverless · Microservice · AWS · Performance · Cost · Case Study

1 Introduction

Microservice architecture is one of the most dominantly popular architectures for cloud-native systems in the industry, where many giant technology companies, e.g., Amazon, LinkedIn, Netflix, and Spotify, are all adopting such an architecture in their systems. In academia, studies on microservice architecture are increasing exponentially in recent years. Many studies focus on the different critical aspects regarding the architectural best practices and issues of microservice architecture, including, the patterns and anti-patterns [27,28], the decomposition of monolithic systems towards microservices [30], the technical debt of monolithic system migration [20], and so on. Many studies also contribute to the systematic analysis of microservice architecture from different application layers in order to support its monitoring and maintenance, e.g. the architecture

H. Han and E. Baker (Eds.): SDSC 2023, CCIS 2113, pp. 60–72, 2024.
https://doi.org/10.1007/978-3-031-61816-1_5

reconstruction and visualization techniques [8], static analysis techniques for architectural reconstruction [7], reconstruction visualization [9], and etc.

Compared to microservice, serverless is an emerging technology that enables the reduction of unnecessary overhead for provisioning, scaling, and general infrastructure management [26]. The industry has also seen such benefits and started the migration to the new paradigm of serverless [29]. Many studies also contribute to the theoretical foundation concerning the many aspects of serverless computing, including the patterns and anti-patterns for serverless functions [26], the economic and architectural impact [1], the design and implementation [24], potential issues and solutions [25] and the even broader application of serverless edge computing [3] as well as the platforms for such a purpose [13].

Despite the benefits of serverless mentioned above, many practitioners and companies are still not certain whether being beneficial to migrate from microservice to serverless. The performance and the cost are two of the main concerns without proper addressing. Many studies have contributed to the analysis and modeling of the performance of either microservice [17] or serverless [23]. Several studies also contribute to the comparison of microservice architecture performance and that of monolithic systems [2,6,15]. Regarding the cost, studies contributed to the analysis of AWS billing estimation as well as the prediction and optimization of the cost of serverless workflow [12,14]. However, the direct comparison studies on microservice and serverless are limited.

In this study, we compare the performance of containerized microservice and serverless functions in terms of response time via an experiment using an identical demo system. The system is designed with identical three-service architecture and ran on both AWS EC2 (microservice) and AWS Lambda (serverless). We use Locust open-source load testing tool to test and compare the performance of both parties in different loading volumes. We also analyze the cost of both configuration plans and identify a potential threshold where the cost comparison may switch.

The contribution of this paper addresses the following research questions (RQs):

RQ1. What is the difference in response time between microservice and serverless systems?

RQ2. What is the difference in cost of running microservice and serverless systems on AWS environments?

The rest of the article is organized as follows: Sect. 2 provides information about related work on the performance and cost of microservices and serverless architecture, including comparative studies. Section 3 explains the experiment system's design and analysis methods in detail. The experiment's results are presented in Sect. 4. Section 5 discusses potential limitations, future work, and validity threats. Finally, Sect. 6 concludes the article.

2   Related Work

Performance engineering for microservices, in terms of testing, monitoring, and modeling, is one of the challenges in the related domain [16]. Especially regarding performance testing, many approaches have been proposed. For example, De Camargo et al. [10] proposed an automatic testing method where each microservice shall provide a test specification. Regarding the performance comparison between microservices and monolithic systems, Auer et al. [4] proposed an assessment framework that encompasses the measures of function suitability, performance efficiency, reliability, maintainability, process-related, and cost as the key dimensions. Regarding the performance comparison between microservice and monolithic systems, Blinowski et al. [6] conducted a series of controlled experiments in different deployment environments to verify the different benefits of the migration from monolithic systems to microservice in various context settings. Gos and Zabierowski [15] also conducted experiments comparing the performance of microservice and monolithic systems in terms of response time for different request numbers and indicating the pros and cons of both architectures. Al-Debagy and Martinek [2] drove a comparison experiment on load testing and service discovery scenarios with specific configurations between the two architectures.

Testing serverless applications is a complex task. While different tools might support the testing process, integration testing still remains an open issue [19,21].

On the other hand, many studies also contributed to the performance engineering regarding serverless applications in terms of the various quality aspects of the architecture. Eismann et al. [11] conducted a case study towards investigating the stability of performance tests for serverless applications by comparing the results with different load levels and memory sizes. Their findings show improvement in the response time (faster responses) with higher workloads and also larger function sizes, as well as performance fluctuations in the short-term and long-term within the observation period. Lloyd et al. [22] investigate the influencing factors of infrastructure elasticity, load balancing, provisioning variation, infrastructure retention, and memory reservation size by comparing such attributes of AWS Lambda and Azure Functions. Their results indicate that extra infrastructure is provisioned to compensate for the initialization overhead of COLD service requests. Lee et al. [18] compared and evaluated concurrent invocations on Amazon Lambda, Microsoft Azure Functions, Google Cloud Functions, and IBM Cloud Functions. Their results show that the elasticity of Amazon Lambda outperforms the others regarding throughput, CPU performance, network bandwidth, and file I/O in terms of concurrent function invocations for dynamic workloads. Yu et al. [33] proposed a benchmark suite for characterizing serverless platforms and compared the evaluation results on AWS Lambda, Apache OpenWhisk, and Fn serverless platforms.

Despite the studies in performance engineering for either microservice or serverless systems, limited contributions to the performance comparison between them. Fan et al. [14] performed a performance comparison study of a cloud-native application regarding its reliability, scalability, cost, and latency between microservices and serverless strategies. They conducted experiments using an employee time-sheet management system developed with Node.js with three main modules when their deployment strategy is 5 containers for 5 main cases for microservices and 6 Lambda functions for serverless. The results show that serverless suffers from cold-start issues and is outperformed by microservice with small size and repetitive requests. Their results also show that microservices suffer from the load balancing and traffic redistribution problem, nevertheless, they do not provide performance comparison with different request load numbers.

Regarding the cost comparison, Fan et al. [14] use AWS billing estimation and the pricing model of AWS ECS and Lambda as references; they compare the costs based on the number of requests received to the functions and the time it takes for the code to execute. However, the study does not take into account the concurrent load as an influencing factor. Furthermore, several other studies investigate the cost of serverless and microservice systems. Eismann et al. [12] provided a method to predict the cost of serverless workflow which supports the prediction and optimization of the expected cost of a serverless workflow.

3 Study Design

In this section, we detail the design of our study that aims to compare the performance of microservice and serverless systems. Specifically, we will compare microservices' implemented using Docker containers deployed on AWS EC2 instances, with microservices deployed as serverless functions on AWS Lambda.

Both systems are versions of the same application, with one being a Dockerized microservice container and the other being a collection of Lambda functions. Each system performs the same functionality as an individual service and will be exposed using the same API gateway instance.

The study has established certain metrics to assess the outcome, which are also aligned with the RQs of the article:

- **Response time** measured in milliseconds for each functionality in the system during tests, given a period of time.
- **Cost** for running both systems on AWS.

3.1 Experiment Design

To compare microservice and serverless applications and answer the questions, we chose a microservice-based application that can be deployed on docker containers and transformed into a serverless-based application. The chosen application must be built with a microservice architecture that can be divided into independent serverless functions and written in a programming language compatible with the AWS Lambda environment among the following:

- Node.js 16/14/12
- Python 3.9/3.8/3.7
- Java 11/8
- Ruby 2.7
- .NET Core 3.1/.NET 6/.NET 5
- Go 1.x

We selected the Bookshop application[1] as the testbench of this study. The Bookshop application is a demonstration of microservices and consists of three separate services

---

[1] https://github.com/happy-bhesdadiya/microservices-demo.

that track customer information, book information, and order details including information about which customer bought which book. These services are written in NodeJS using the express framework and interact with the same MongoDB database. The order service retrieves information from the other two services using http method calls.

The EC2 deployed version is depicted in Fig. 1a. It consists of three loosely-coupled microservices that operate within the same docker network and can send requests to one another, with external communication going through ports that are exposed via the EC2 instance's public IPv4 address. The instance is also connected to Amazon Cloudwatch for metrics and an S3 bucket for storage.

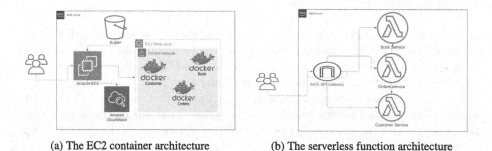

(a) The EC2 container architecture   (b) The serverless function architecture

Fig. 1. Architecture Comparison between AWS EC2 and Lambda

The Lambda application is made of three different Lambda functions that each represent one of the microservices from the original application. Each of them is exposed externally through endpoints routed through an instance of AWS API gateway, as illustrated in Fig. 1a. They communicate directly by using execution links. Each serverless code excerpt was deployed on a separate Lambda instance that was running on a NodeJS 12.x run-time. Serverless is not compatible with ExpressJS, which is the used framework in the original application, so a JS package called Serverless Express is used to wrap each function in a handler.

In order to assess the performance of both systems, unit tests were conducted on both the microservice and serverless versions using Locust.py to simulate mock user requests. The tests were performed with 10 concurrent users first, then increased to 50, 100, 500, and finally 1000, each for approximately 30 s. Although the tests were virtually identical, they had different endpoints due to the differences between the systems.

During the test, each simulated user was lined up at the start, and they frequently made http requests with different weights. When a new entry was created in the MongoDB persistence storage during the POST actions, random strings were generated, and when querying the ID of the returned items from the GET requests, the IDs were stored and then accessed. The requests were made in a random order.

3.2   Data Analysis

The Locust tool was used to collect execution and performance data for the two systems under test. The tool was run on a local machine, and the URL of each system was passed to the tool. The tool created the specified number of users at a rate of x users per second until the desired number was reached. The users continually executed tasks specified in a text file, and the tool generated a table and graph in HTML format with raw data in CSV format after the tests were manually stopped.

To analyze the results, the data generated by the tool was plotted on a graph at different intervals (5, 10, 15, 20, 25, and 30 s) to compare the performance of the two systems and identify trends as the number of concurrent users increased over time.

The response time and the cost of operating both systems are compared against each other. While the response time is measured in milliseconds, the cost is calculated by applying the number of executions to the individual cost per execution. The cost of the EC2 instance is defined per on-demand instance hour. The precise number of function requests made and computing hours used is taken from AWS' billing center for the month when the tests were carried out. The Amazon billing tool was used to obtain the total number of times the Lambda functions were triggered in a month.

4   Results

Herein, we present the results towards answering the research questions in terms of the performance and cost comparison between AWS EC2 (microservice) and AWS Lambda (serverless).

4.1   RQ1. What Is the Difference in Response Time?

In order to compare the performance of the two architectural styles, we first simulate 10 concurrent users using the bookstore application in both AWS Lambda and AWS EC2. We run the simulation 10 times in order to avoid potential external influences. For each round of the simulation, we record the response time at 5, 10, 15, 20, 25, and 30 s respectively. The result is shown in Fig. 2.

We can observe that, regarding the medians of both parties, their performances in terms of response time are very similar, and both are under 1000 milliseconds, although the performance of EC2 is slightly better than that of Lambda in the range of 0 to 15 s. However, an interesting phenomenon is that, regarding the 95th Percentile, the response time of Lambda is more than six times longer than that of EC2 in the range of 5 to 10 s. It means there are 5% of the users shall experience a 7-second long lagging when sending a request to Lamdba. However, when the application has been started for longer ($\geq$ 25 s) the 95th percentile performance of Lamdba improves to nearly the level of EC2.

Furthermore, we also conduct the same simulation experiments for 50, 100, 500, and 1000 concurrent users in order to verify the consistency of the previous comparison. The comparative results are shown in Fig. 3. In terms of the performance median, we can easily observe that Lambda often has a longer response time in the beginning (5 - 15 s) than EC2 and then improve to a similar level of EC2 after this period. Such a phenomenon does not stand when the user number increases up to 500, where the performance of Lamdba exceeds that of EC2 by more than 10 to 30 times after the first 20 s "cold start".

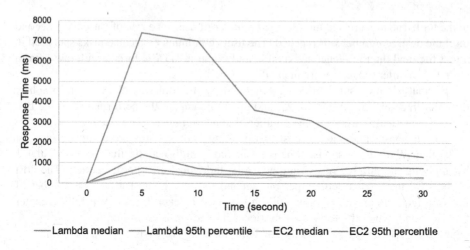

Fig. 2. Performance Comparision between AWS EC2 and Lambda (10 Concurrent Users)

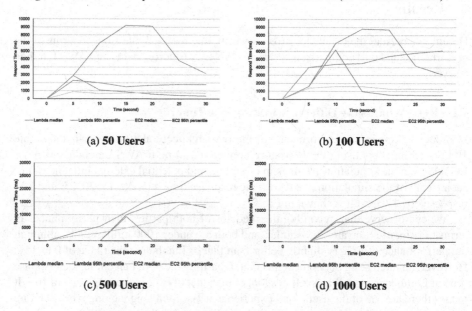

Fig. 3. Performance Comparison with Different User Numbers

On the other hand, regarding the 95th percentile performance with 50 and 100 concurrent users, the significant lagging phenomenon for Lambda persists for the first 20 s. The maximum response time can reach 9000 milliseconds. However, similar to the situation with 10 users, the 95th percentile performance of Lambda starts to improve after the 20 s "cold start", but still cannot reach the level of EC2. To be noted, starting from 100 concurrent users, the 95th percentile performance of EC2 is getting worse when such performance deterioration becomes even more severe as the user number grows.

Especially, with more than 500 concurrent users, the 95th percentile response time of EC2 grows almost exponentially with the experiment time.

4.2 RQ2. What Is the Difference in Cost?

Regarding the costs of the two architectural setups, we compare the potential cost when holding the same amount of requests frequency. Meanwhile, we also compare the cost difference in terms of different request rates.

Firstly, we calculate the potential monthly cost of EC2 service by taking into account the less costly example solutions. By exploring the "Savings Plans"[2], we select the region of "US EAST (N. Virginia)" with the shared-tendency Linux operating system with the payment options of 1-year term length and "no upfront". For such a basic configuration, the on-demand hourly cost rate for a "t2.micro" instance (1 vCPUs, 1 GB memory, Low to Moderate network performance, EBS storage) is 0.0116 USD. Therefore, the basic monthly payment for such an EC2 instance shall be $0.0116 \times 24 \times 30 = 8.352$ USD/month.

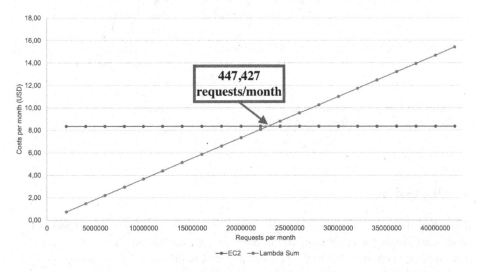

Fig. 4. Costs Comparison of AWS EC2 and Lambda

On the other hand, we calculate the cost of AWS Lambda service by taking into account the same configuration. We adopt the "US EAST (N. Virginia)" region as the default setting with which the monthly compute price is $1.67e-5$ USD/GB-s. With the Lambda function executed n times per month and running for 10 ms each time, the monthly compute charges will be $n \times 1.67e - 5 \times 0.01$. Meanwhile, the basic monthly request price is 0.2 USD per million requests for the starting 6 billion GB-s month. Therefore the monthly request charges shall be calculated as $n \times 0.2 \times 1e-6$. Therefore,

---

[2] https://aws.amazon.com/savingsplans/compute-pricing.

the total Lambda function monthly charge is the sum of the compute charge and the request charge when the cost has a linear relation to the number of monthly requests. These calculations are summarized in Table 1.

As shown in Fig. 4, we can observe the relatedness of the monthly costs of both EC2 and Lambda where the intersection point can be easily calculated. For a simplified application scenario like the above-mentioned configuration, when a service receives no more than 447427 requests per month, Lambda is a better option for savings.

Table 1. Cost Calculation for EC2 Vs Lambda

	EC2	Lambda
Measurement Unit	number hours per month	execution duration per month
Estimation Formula (per month)	hourly cost × #hours per day × #days per month	#executions × price per request
Cost for Basic Configurations	$0.0116 \times 24 \times 30 =$ $8.352/month$	$n \times (0.2 \times 1e - 6)$; n is the number of requests per month
Calculation Tool	AWS Billing center[a]	AWS Billing center/dashbird.io Lambda cost calculator[b]

[a]https://aws.amazon.com/aws-cost-management/aws-cost-explorer.
[b]https://dashbird.io/lambda-cost-calculator.

5 Discussion

Regarding the response time comparison outcomes (RQ1), we can easily observe the "cold start" issue for AWS Lambda in terms of both the median and 95th Percentile. Such a phenomenon is due to the fact that the first request for a new Lambda worker needs to find a space in the EC2 fleet to allocate and initialize. According to an analysis of production Lambda workloads provided by AWS Compute Blog, cold starts typically occur in under 1% of invocations[3]. Such an inference can be seen as supported by our outcomes showing that the "cold start" issue for the 95th percentile is more obvious and somehow insufferable due to the lagging experience. Several studies have provided potential solutions for solving the "cold start" issue of Lambda, e.g., by reducing container preparation and function loading delay, invoking function periodically preventing cold functions, using application knowledge on the function composition, ... etc. [5, 31]. Furthermore, it is also noticeable that with a growing number of concurrent users, the 95 percentile performance deteriorates significantly for both EC2 and Lambda. The reason

---

[3] https://aws.amazon.com/blogs/compute/operating-lambda-performance-optimization-part-1.

is likely due to the selected experiment configuration with only limited resources allocated. With a larger, more complex system the expectation would be that the difference in performance between the two systems would remain largely the same. The limitation shall be addressed in future studies by using a higher level of configuration settings and testing with a larger number of concurrent users. Another potential limitation lies in the internal performance of the two systems which is not factored in when comparing the performance based on response times. In future works, there could be a model devised to measure what impact the time for services to communicate with one another has.

On the other hand, regarding the cost comparison (RQ2), we adopt a basic application scenario for both EC2 and Lambda and find that when the number of requests per month is under a certain level (e.g. in this scenario, 447,427) using Lambda is a more cost-saving option. Due to the various "saving plans" and configuration options provided by AWS, the "sweet spot" for switching services shall inevitably vary. Considering large search engine companies process about 400k searches in about 10 s, their preferred option shall still be EC2 with a fixed monthly quote instead of Lambda, though the eventual number will largely exceed the number for our bookstore demo scenario. To such an end, the future shall be conducted towards a more comprehensive cost calculation model with a set of critical cost parameters taken into account.

5.1   Threats to Validity

This section addresses the threats to the validity of our research. We consider Wohlin's taxonomy [32] for this.

Construct Validity: We have implemented two system versions, one for each approach. We used development frameworks that are used in the enterprise architecture to develop such systems, each system with comparable resources. Moreover, we used conventional practices aligning with the particular approach to developing these systems. Regarding the system size, it is limited but sufficient to demonstrate the differences. The simulation traffic has been fabricated and reflects conventional system testing approaches. To measure performance, we used the established tool, locust.py, to mitigate inaccurate methods. The format of testing and measurement could present a validity threat, however, we used established practices and tools to mitigate these construct threats.

Internal Validity: The first potential threat to internal validity is related to the fact that one author developed both system versions to ensure that they have the same functionality and comparable amount of computing resources. There may be some unintentional biases or errors in the development process that could affect the results. However, other authors verified the implementation. Another potential threat is related to the timing deviation in the measurements. To limit this, the experiment was repeated 10 times for 30 s each, and the values were averaged.

External Validity: Regarding the case study, we have used a small system benchmark; however, all the design principles remain the same for arbitrary sizes. The performance evaluation must be assumed in the context of a small system limiting the perspective on sample data access operations, not involving complex business logic or data routings. The motivation of the study is not to derive exact costs but to draw the relative difference between the two considered approaches, and the findings render themselves significant.

Conclusion Validity: We minimized the risk of author bias in the study by eliminating any potential mechanisms that could influence the performance of the two cloud models. Furthermore, we employed a small system with two implementations to compare the performance and costs of two cloud design approaches. As a result, our study findings demonstrate a substantial cost difference and a noticeable performance variation as the number of users increases.

6   Conclusion

The purpose of this paper was to take a new technology growing in popularity and compare it against an existing one, measuring both cost and response time. By comparing the performance of the same application deployed on both AWS EC2 and Lambda, we can observe a significant low 95th percentile performance for Lambda at the starting phase. Meanwhile, the 95th percentile performance for EC2 deteriorates more significantly than Lambda with an increasing concurrent user number. Regarding the cost of both parties, it is more expensive to adopt EC2 with limited requests frequency, when as the monthly requests grow greater than a certain threshold the price of Lambda will be increasingly higher than EC2. For future work, more testing scenarios and parameters shall be taken into account regarding the responsive performance and a comprehensive cost model shall also be investigated with multiple extra pricing factors considered.

References

1. Adzic, G., Chatley, R.: Serverless computing: economic and architectural impact. In: Proceedings of the 2017 11th Joint Meeting on Foundations of Software Engineering, pp. 884–889 (2017)
2. Al-Debagy, O., Martinek, P.: A comparative review of microservices and monolithic architectures. In: 2018 IEEE 18th International Symposium on Computational Intelligence and Informatics (CINTI), pp. 000149–000154. IEEE (2018)
3. Aslanpour, M.S., et al.: Serverless edge computing: vision and challenges. In: 2021 Australasian Computer Science Week Multiconference, pp. 1–10 (2021)
4. Auer, F., Lenarduzzi, V., Felderer, M., Taibi, D.: From monolithic systems to microservices: an assessment framework. Inf. Softw. Technol. **137**, 106600 (2021)
5. Bermbach, D., Karakaya, A.S., Buchholz, S.: Using application knowledge to reduce cold starts in faas services. In: Proceedings of the 35th Annual ACM Symposium on Applied Computing, pp. 134–143 (2020)
6. Blinowski, G., Ojdowska, A., Przybyłek, A.: Monolithic vs. microservice architecture: a performance and scalability evaluation. IEEE Access **10**, 20357–20374 (2022)
7. Bushong, V., Das, D., Cerný, T.: Reconstructing the holistic architecture of microservice systems using static analysis. In: CLOSER, pp. 149–157 (2022)

8. Cerny, T., Abdelfattah, A.S., Bushong, V., Al Maruf, A., Taibi, D.: Microservice architecture reconstruction and visualization techniques: a review. In: 2022 IEEE International Conference on Service-Oriented System Engineering (SOSE), pp. 39–48. IEEE (2022)

9. Cerny, T., Abdelfattah, A.S., Bushong, V., Al Maruf, A., Taibi, D.: Microvision: static analysis-based approach to visualizing microservices in augmented reality. In: 2022 IEEE International Conference on Service-Oriented System Engineering (SOSE), pp. 49–58. IEEE (2022)

10. De Camargo, A., Salvadori, I., Mello, R.D.S., Siqueira, F.: An architecture to automate performance tests on microservices. In: Proceedings of the 18th International Conference on Information Integration and Web-based Applications and Services, pp. 422–429 (2016)

11. Eismann, S., et al.: A case study on the stability of performance tests for serverless applications. J. Syst. Softw. **189**, 111294 (2022)

12. Eismann, S., Grohmann, J., Van Eyk, E., Herbst, N., Kounev, S.: Predicting the costs of serverless workflows. In: Proceedings of the ACM/SPEC International Conference on Performance Engineering, pp. 265–276 (2020)

13. El Ioini, N., Hästbacka, D., Pahl, C., Taibi, D.: Platforms for serverless at the edge: a review. In: Zirpins, C., et al. (eds.) ESOCC 2020. CCIS, vol. 1360, pp. 29–40. Springer, Cham (2021). https://doi.org/10.1007/978-3-030-71906-7_3

14. Fan, C.F., Jindal, A., Gerndt, M.: Microservices vs serverless: a performance comparison on a cloud-native web application. In: CLOSER, pp. 204–215 (2020)

15. Gos, K., Zabierowski, W.: The comparison of microservice and monolithic architecture. In: 2020 IEEE XVIth International Conference on the Perspective Technologies and Methods in MEMS Design (MEMSTECH), pp. 150–153. IEEE (2020)

16. Heinrich, R., et al.: Performance engineering for microservices: research challenges and directions. In: Proceedings of the 8th ACM/SPEC on International Conference on Performance Engineering Companion, pp. 223–226 (2017)

17. Jindal, A., Podolskiy, V., Gerndt, M.: Performance modeling for cloud microservice applications. In: Proceedings of the 2019 ACM/SPEC International Conference on Performance Engineering, pp. 25–32 (2019)

18. Lee, H., Satyam, K., Fox, G.: Evaluation of production serverless computing environments. In: 2018 IEEE 11th International Conference on Cloud Computing (CLOUD), pp. 442–450. IEEE (2018)

19. Lenarduzzi, V., Daly, J., Martini, A., Panichella, S., Tamburri, D.A.: Toward a technical debt conceptualization for serverless computing. IEEE Softw. **38**(1), 40–47 (2021). https://doi.org/10.1109/MS.2020.3030786

20. Lenarduzzi, V., Lomio, F., Saarimäki, N., Taibi, D.: Does migrating a monolithic system to microservices decrease the technical debt? J. Syst. Softw. **169**, 110710 (2020)

21. Lenarduzzi, V., Panichella, A.: Serverless testing: tool vendors' and experts' points of view. IEEE Softw. **38**(1), 54–60 (2021). https://doi.org/10.1109/MS.2020.3030803

22. Lloyd, W., Ramesh, S., Chinthalapati, S., Ly, L., Pallickara, S.: Serverless computing: an investigation of factors influencing microservice performance. In: 2018 IEEE international conference on cloud engineering (IC2E), pp. 159–169. IEEE (2018)

23. Mahmoudi, N., Khazaei, H.: Performance modeling of serverless computing platforms. IEEE Trans. Cloud Comput. **10**(4), 2834–2847 (2020)

24. McGrath, G., Brenner, P.R.: Serverless computing: design, implementation, and performance. In: 2017 IEEE 37th International Conference on Distributed Computing Systems Workshops (ICDCSW), pp. 405–410. IEEE (2017)

25. Nupponen, J., Taibi, D.: Serverless: what it is, what to do and what not to do. In: 2020 IEEE International Conference on Software Architecture Companion (ICSA-C), pp. 49–50. IEEE (2020)

26. Taibi, D., El Ioini, N., Pahl, C., Niederkofler, J.R.S.: Patterns for serverless functions (function-as-a-service): a multivocal literature review (2020)
27. Taibi, D., Lenarduzzi, V., Pahl, C.: Architectural patterns for microservices: a systematic mapping study. In: CLOSER 2018: Proceedings of the 8th International Conference on Cloud Computing and Services Science, Funchal, Madeira, Portugal, 19–21 March 2018. SciTePress (2018)
28. Taibi, D., Lenarduzzi, V., Pahl, C.: Microservices anti-patterns: a taxonomy. Microserv. Sci. Eng. 111–128 (2020)
29. Taibi, D., Spillner, J., Wawruch, K.: Serverless computing-where are we now, and where are we heading? IEEE Softw. **38**(1), 25–31 (2020)
30. Taibi, D., Systä, K.: From monolithic systems to microservices: a decomposition framework based on process mining (2019)
31. Vahidinia, P., Farahani, B., Aliee, F.S.: Cold start in serverless computing: current trends and mitigation strategies. In: 2020 International Conference on Omni-layer Intelligent Systems (COINS), pp. 1–7. IEEE (2020)
32. Wohlin, C., Runeson, P., Hst, M., Ohlsson, M.C., Regnell, B., Wessln, A.: Experimentation in Software Engineering. Springer, Heidelberg (2012). https://doi.org/10.1007/978-3-642-29044-2
33. Yu, T., et al.: Characterizing serverless platforms with serverlessbench. In: Proceedings of the 11th ACM Symposium on Cloud Computing, pp. 30–44 (2020)

NBD-Tree: Neural Bounded Deformation Tree for Collision Culling of Deformable Objects

Ryan S. Zesch[1]([✉]), Bethany R. Witemeyer[1], Ziyan Xiong[1],
David I. W. Levin[2], and Shinjiro Sueda[1]

[1] Department of Computer Science and Engineering, Texas A&M University,
College Station, TX, USA
rzesch@tamu.edu
[2] Department of Computer Science, University of Toronto, Toronto, ON, Canada

Abstract. We propose a novel machine learning-based approach for accelerating the broad phase of 3D collision detection for deformable objects. Our method, which we call the neural bounded deformation tree (NBD-Tree), allows us to cull away primitives for full-space deformable objects quickly. Unlike its classic, non-neural counterpart, the NBD-Tree is not limited to deformable objects that are constrained to work within the space of low-dimensional deformation modes, and instead works with an arbitrary set of deformations. With our approach, when the shape of the object changes at runtime, we use the low-dimensional deformation modes of the object only as the input to a neural network that calculates the necessary updates to the NBD-Tree. To further improve efficiency, we approximate these low-dimensional modes efficiently through clustering, which allows us to avoid going through every vertex of the mesh. We then rely on the network to overcome the potential errors stemming from these approximations. The NBD-Tree paves the way for interactive collision culling of large-scale, full-space deformable objects.

Keywords: Collision Detection · Neural Network · Broad Phase · Bounding Volume Hierarchy · Sphere Tree

1 Introduction

3D collision detection is an important component of various applications in such fields as computer graphics, computer vision, and robotics, but an efficient approach for deformable objects remains a challenge [22]. In most applications, the collision detection pipeline is usually divided into two parts: the broad-phase to quickly cull away primitives that are far from being in contact, and the narrow-phase to go through the remaining primitives to compute the actual collisions [4]. In this paper, we focus on broad-phase collision detection. In particular, we work with a sphere tree, which is one of the most commonly-used techniques for broad-phase collision detection. With a sphere tree, we surround the 3D object

H. Han and E. Baker (Eds.): SDSC 2023, CCIS 2113, pp. 73–87, 2024.
https://doi.org/10.1007/978-3-031-61816-1_6

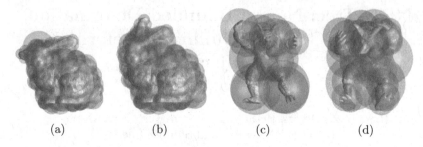

(a)   (b)   (c)   (d)

Fig. 1. The Neural Bounded Deformation Tree allows us to quickly compute a sphere tree for full-space deformable objects.

with a hierarchical set of spheres. If the root sphere of a 3D object, which contains all of the primitives of the 3D object, does not intersect the root sphere of another 3D object, then we know that the two 3D objects are not intersecting; otherwise, we recursively check the children spheres of the two sphere trees.

For objects that are constrained to deform linearly based on a set of modes [21], the classic bounded deformation Tree algorithm (BD-Tree) is still arguably the best choice for the broad-phase [9]. The BD-Tree algorithm starts with a sphere tree built from the rest pose of the deformable object, and then during runtime, it updates the sphere centers and radii in time linear in the number of modes, regardless of the number of vertices/triangles in the mesh. In some situations, however, using the full space, as opposed to the linear modes, of deformations is desired or necessary, especially when hard constraints are present, as they can cause locking with reduced deformations.

Unfortunately, BD-Trees cannot be used for full-space deformations, since its update equations depend on the linearity of the deformations—large full-space deformations cause BD-Tree bounds to be extremely conservative. Therefore, we propose the neural bounded deformation tree (NBD-Tree), which uses neural networks to update the sphere tree for full-space deformable objects (Fig. 1). At runtime, we compute the low-dimensional modes of the deformation, which are passed through a network to compute the corrections to be applied to the rest pose sphere tree. We use a small multi-layer perceptron (MLP) for each sphere of the tree, making the network evaluation very fast. However, the process of computing the low-dimensional modes can then become a bottleneck for a large mesh because: first, we need to compute the rigid alignment of the mesh to go to the local transformed space of the trained network, and then we need to perform a matrix-vector multiplication between the modal matrix and the transformed vertices to compute the low-dimensional modes. Unfortunately, both of these operations are linear, requiring us to go through every vertex of the mesh. Therefore, we compute the rigid alignment and the matrix-vector multiplication via a clustering approach to quickly compute the approximate low-dimensional mode, and then rely on the network to overcome the potential errors stemming from the approximations. This allows us to compute the modes quickly, thereby making the whole pipeline highly efficient.

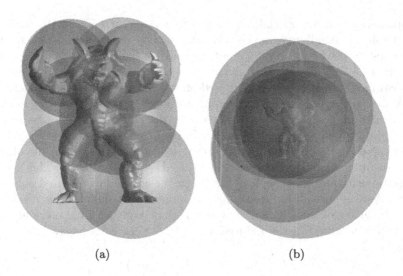

(a)   (b)

Fig. 2. Even at a mild deformation (b) from the rest pose (a), a standard BD-Tree is vastly overly conservative when using a small number of modes of deformation.

2   Related Work

3D collision detection has been an active area of research in many fields including graphics, robotics, and vision, with several survey papers spanning multiple decades [8,11,22]. We refer the reader to these excellent surveys for an overview of various techniques. One of the most popular approaches for deformable objects collisions is a bounding volume hierarchy (BVH). If the modes of deformation are known *a priori*, then the individual bounding volumes in a BVH can be updated very efficiently [9]. However, as mentioned in the introduction, these fast updates do not work on full-space models. Image-based methods work well with deformable objects and naturally run on the GPU [5,23], but these methods cannot be readily incorporated into other simulation frameworks. Another approach is to deform a signed distance field (SDF) based on the object's mesh [6,12,13]. However, with these methods, a BVH is still required to find the region or the cell that contains the query point. Deformed SDFs have also been used for deformed sphere tracing and simple collision detection [20], but such methods have limited applicability to general collision detection because they cannot evaluate the underlying implicit surface at an arbitrary point in deformed space.

Recently, approaches based on neural fields have become extremely popular [24]. Of these, implicit shape representation through occupancy or signed distance fields are highly relevant to collision detection. Park et al. [16] showed that, with their Coded Shape DeepSDF approach, they can build a highly effective implicit representation of non-rigid 3D geometry. Concurrent work by Mescheder et al. [14] and Chen and Zhang [2] used neural networks for occupancy fields. All of these works use neural approaches for various visual applications, such as shape completion, interpolation, and 3D reconstruction. There are also neural approaches that are specialized for articulated characters [1,3,19], including implicit collision handling with posed characters. Our work is orthogonal to these approaches, focused on general volumetric collision handling.

In the BD-Tree algorithm [9], which we base our work upon, a sphere tree for a deformable model is updated based on its current reduced space deformation. A rest-pose sphere tree is precomputed using a wrapped hierarchy, in which each sphere contains the enclosed geometry of its children, but not necessarily the bounding spheres of its children. At runtime, a sphere tree node has its center and radius updated using precomputed quantities derived from the model's displacement field and the current set of reduced space coordinates. This method is output-sensitive in that a sphere has its center and radius updated only if its parent sphere is in collision, reducing the number of computations needed. A central drawback of this method, however, is that large deformations and deformations not captured in the reduced space have very conservative bounds, as seen in Fig. 2.

3   NBD-Tree Algorithm

3.1   BD-Tree Overview

With a standard sphere tree, whenever the 3D object changes its shape, the sphere centers and radii are updated by checking all the primitives of the 3D object. If we constrain the deformation to only linear modes, we can instead use the Bounded Deformation Tree (BD-Tree) algorithm [9]. This work is based upon the fact that a deformed model $x \in \mathbb{R}^{3n}$ which has rest pose $X \in \mathbb{R}^{3n}$ can be approximated as $x = X + Uq$, where $U \in \mathbb{R}^{3n \times m}$ is the model's displacement field and $q \in \mathbb{R}^m$ is a vector of reduced space coordinates (i.e., linear modes) for the current deformation. In the BD-Tree, each node of a sphere tree can be updated to the model's current deformation independently by using precomputed values for that model along with the linear modes for the current deformation. In particular, a sphere's center and radius are updated as $c' = c + \bar{U}q$ and $r' = r + \Delta r^T \mathrm{abs}(q)$, where $\bar{U}$ and Δr are precomputed matrices derived from U. The updated spheres computed via these matrices are guaranteed to contain the deformed model, but if deformations are not well approximated by U, BD-Tree bounds become extremely conservative. In our method, we address these overly conservative bounds by using a learning based approach, which we call the Neural Bounded Deformation Tree (NBD-Tree).

3.2   NBD-Tree Overview

An overview of the online portion of the NBD-Tree pipeline of is shown in Fig. 3a. At every frame of the simulation, given the current vertex positions x, our goal is to update the predicted (binary) sphere tree $\bar{S}_x$. (We use a bar above S to indicate that this is the *predicted* quantity computed by evaluating the network.) This is done through a sequence of steps, including rigid alignment ($\hat{y}$), code

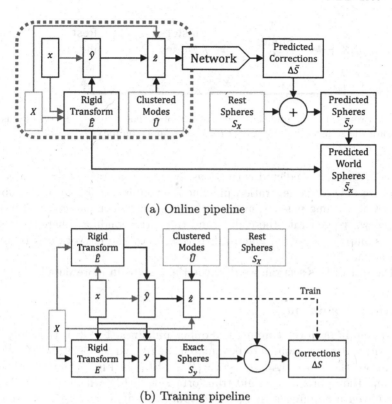

(a) Online pipeline

(b) Training pipeline

Fig. 3. Training and online pipelines. Green boxes are the quantities precomputed in the preprocessing pipeline, shown in Fig. 4. Red arrows imply that a subset of the quantity is used to generate a reduced quantity. In the online pipeline, the portion inside the dashed blue rectangle is performed once for the entire tree, whereas the other portions are evaluated multiple times to update the spheres as necessary. (Color figure online)

generation ($\hat{z}$), and network evaluations. (We use the hat notation to indicate quantities that result from clustering, which we describe later.) The output of the network is a correction $\Delta \bar{S}$ to be applied to S_X, the sphere tree created during the preprocessing stage with the rest pose vertex positions X, shown in Fig. 4. We follow the original BD-Tree approach of using a *wrapped*, instead of *layered*, hierarchy [7,9]. Like the BD-Tree algorithm, our method is output-sensitive in that we update the sphere tree nodes only if necessary. For instance, if the top-level sphere does not return a collision, we do not update any of the descendant spheres. In Fig. 3a, this is indicated by the dotted rectangle in dark blue. The portion of the pipeline within this rectangle is performed once for the entire tree, whereas the remaining portions are evaluated for a sphere if its parent sphere reported a collision.

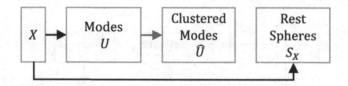

Fig. 4. Preprocessing pipeline. The red arrow implies that a subset of rows of U is used to generate the reduced modal matrix $\hat{U}$. (Color figure online)

To compute the predicted corrections, we use a neural network trained for each sphere. The data generation pipeline for training is shown in Fig. 3b. We take a set of training poses x, and, following a series of steps described in more detail below, we generate the code $\hat{z}$ and the corresponding sphere corrections ΔS. This mapping between $\hat{z}$ and ΔS is learned by the network and is used in the online pipeline.

In the rest of this section, we describe these steps in more detail.

3.3   Rigid Alignment

The first step in both the training and online pipelines (Fig. 3) is to rigidly align the current vertex positions x to best match the rest vertex positions X. We will first consider the exact rigid transform $E \in SE(3)$ in Fig. 3a, ignoring the red arrows and the approximate rigid transform $\hat{E} \in SE(3)$, which we will describe shortly. Given a training pose x, we compute the rigid transform E using the method described by Müller et al. [15]. This rigid transform then allows us to compute the aligned vertex positions:

$$y = Ex. \tag{1}$$

Using these aligned vertices, we construct the exact spheres S_y in the local aligned space. Using the local space is helpful for training the network, since then the network would not need to learn the rigid transforms.

For a large mesh, computing the best rigid transform can become a bottleneck, since we must go through every vertex of the mesh. We therefore use a pre-selected subset of vertices to efficiently compute the approximate rigid transform $\hat{E}$. (In Figs. 3 and 4, we indicate all of the steps that use the pre-selected subset with red arrows. All the arrows share the same subset.) We then use the same subset of vertices from x to form $\hat{x}$ and transform them by $\hat{E}$:

$$\hat{y} = \hat{E}\hat{x}. \tag{2}$$

In the next subsection, we discuss how we choose this subset of vertices.

3.4   Clustered Modes

Along with the rest sphere tree S_X, we construct the modal matrix U in the preprocessing stage (Fig. 4). This matrix can be constructed in a number of ways, as described by Sifakis and Barbic [21], but in our implementation, we use linear modes based on the mass and stiffness matrices of the volumetric object [18]. The modal basis matrix U, regardless of how it was constructed, is a tall and skinny matrix; if there are n vertices, then U is a $3n \times m$ matrix, where $m \ll n$. (In our implementation, we use $m = 128$ columns.) We can use $U^\top$ to transform from the (aligned) full space y to the modal space z:

$$z = U^\top(y - X). \tag{3}$$

Unfortunately, these operations again require us to iterate over all of the vertices.

Therefore, as a preprocessing step, we cluster the rows of U to form a clustered modal matrix $\hat{U}$ (red arrow in Fig. 4). First, we reshape U so that a single row corresponds to a 3D vertex, rather than a single coordinate, which makes U be $n \times 3m$. We then cluster the rows of this reshaped U via k-means using the standard L^2 metric. After clustering, we find the representative vertex in each cluster that is closest to each centroid. The indices of these k representative vertices become the *pre-selected subset* for efficiently computing the rigid alignment and the modes. Taking the rows corresponding to these k vertices and reshaping, we form the $3k \times m$ matrix $\hat{U}$. To compute the clustered modal space vector $\hat{z}$, we apply a per-cluster weight before multiplying by $\hat{U}^\top$:

$$\hat{z} = \hat{U}^\top \hat{w} \odot (\hat{y} - \hat{X}), \tag{4}$$

where $\hat{w}$ is a weight vector composed of the number of elements in each cluster, and $\odot$ denotes a component-wise multiplication between two vectors.

Since the result of k-means depends on initialization, we run the clustering algorithm multiple times and choose the result that gives the smallest average L^2 distance between z and $\hat{z}$ across all training poses. The inevitable error that comes from the clustered alignment and modes will be remedied by the network, which we describe next.

3.5   Training

The training poses are generated by running an FEM simulation of the volumetric object with various initial and dynamic conditions. As shown in Fig. 3b, for each pose x, we compute the full space rigid alignment y, the clustered rigid alignment $\hat{y}$, and the code $\hat{z}$ as described in Sect. 3.3 and Sect. 3.4. For each tree node, we compute the target bounding sphere $S_y = \{c_y, r_y\}$ around y, where the sphere is defined by its center c_y and radius r_y. We then subtract the rest sphere S_X from S_y to compute the corrections ΔS:

$$\Delta c = c_y - c_X, \quad \Delta r = r_y - r_X. \tag{5}$$

Our neural network will learn a mapping from the clustered modal codes to the sphere corrections: $\hat{z} \mapsto (\Delta c, \Delta r)$. This allows us to quickly reconstruct the sphere without having to visit every vertex of the mesh.

As a preprocessing step, we normalize the range of $\hat{z}, \Delta c$, and Δr to ± 1 element-wise. For each tree node, we train a small MLP. Each MLP uses two

Table 1. Meshes used for our experiments.

Mesh	# Nodes	# Faces	# train	# test
BUNNY	5988	6244	10k	2.7k
ARMA	10518	16204	10k	2k
ARMAHD	32410	64816	6k	1k
ARMAUHD	129634	259264	850	100

fully connected layers with 32 neurons and ReLU activation functions between layers. We use a simple L^2 loss as our loss function. We train all networks in parallel as a single model, and extract the weights and biases for each tree node as a post-processing step.

3.6 Runtime

The runtime pipeline (Fig. 3a) has two stages: code computation and sphere updates. In each timestep, we first compute the code $\hat{z}$ as described in Sect. 3.3 and Sect. 3.4. Next, we recursively update, starting from the root node. Each node's two children are updated only if the node is itself in collision. Finally, if a leaf node is found to be in collision, it passes its stored list of triangles back to be handled by a narrow phase collision detector.

The update procedure is the same for all spheres. We first normalize $\hat{z}$ component-wise and pass it through the network for the sphere. The output of the network is then unnormalized component-wise, which gives us the predicted corrections $\Delta \bar{S} = \{\Delta \bar{c}, \Delta \bar{r}\}$. These predicted corrections are applied to the rest sphere $S_X = \{c_X, r_X\}$ to compute the predicted sphere $\bar{S}_y$ in the local aligned space:

$$\bar{c}_y = c_X + \Delta\bar{c}, \quad \bar{r}_y = r_X + \Delta\bar{r}. \tag{6}$$

Finally, this sphere is transformed via $\hat{E}$ to give us the world space sphere $\bar{S}_x$.

4   Results

All networks in our pipeline were trained in PyTorch [17] on dual Titan-RTX graphics cards. We use the Adam optimizer with a learning rate of 5×10^{-3} with plateau decay [10]. We found that lower batch sizes yield better results, and used a batch size of 128. All networks are trained to convergence. For each tree node, we use an MLP with 2 hidden layers and 32 neurons per layer. We found the benefits of any network larger than this size to be marginal. For all our experiments, we use 128 linear modes, and in our modal clustering, we use 512 cluster points. All of our online code is written in C++ with Eigen, including MLP evaluations. We test our method on four meshes, as listed in Table 1.

All networks for a tree are trained in parallel with an L^2 loss. We experimented with other loss functions, including a loss which highly penalized when

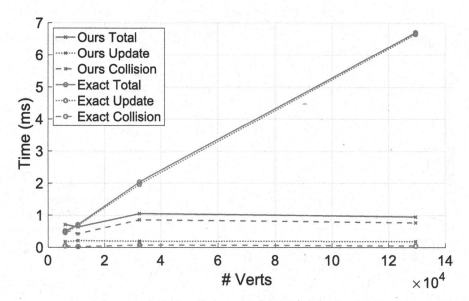

Fig. 5. Comparison of performance between Our method and Exact sphere trees. Total: the amount of time it takes to run the online pipeline for two colliding objects. Update: the amount of time to compute the code by our methods or to update the spheres by the exact. Collision: the amount of time spent computing sphere-sphere collisions. (Color figure online)

predictions do not contain the ground truth nodes. In practice, we found that the performance benefit of such a loss was no better than simply increasing predicted radii by 1%, in terms of percentage of vertices protruding from their predicted nodes.

When we cluster the modal matrix U, we experience a code reconstruction error of at most 5%. Because this error is systematic, our networks are able to overcome it and produce results on par with unclustered methods. We additionally find that increasing the radii of a clustered model by 1–2% will result in vertex containment better than its unclustered counterpart.

Figure 5 shows the timing results for colliding two objects. For these tests, we used a collision scenario that occurs often in practice with physics based modeling: with the two objects touching at a few places but with no deep inter-penetrations. We compare our method against a traditional bottom-up 'exact' sphere tree. With the exact sphere tree, we need to scan the whole mesh to update the tree (red dotted line). On the other hand, with our method (blue dotted line), we are able to quickly compute the code $\hat{z}$ using the clustering approach—no matter how large the mesh is, as long as $\hat{U}$ is the same size, it takes the same amount of time to compute the code (blue rectangle in Fig. 3a). This is a reasonable assumption when working with different resolutions of the same object, since the modes are likely to be similar, and the vertex clusters are also likely to be similar. The actual colliding of the spheres (dashed lines)

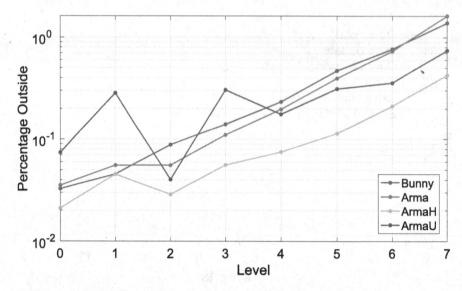

Fig. 6. The percentage of vertices that are outside the spheres at each level (*i.e.*, $10^{-1} = 0.1\%$).

is faster with the exact approach, because once the spheres are constructed, the collision check only involves the cheap distance check between spheres. Our method is required to evaluate an MLP to compute the center and the radius, and so it is relatively more expensive to perform these checks. In essence, the long-term trajectory of our approach does not depend on the vertex count but on the modes and the clusters, all the while being able to account for full-space deformations due to the neural corrections.

We find that our trained networks learn bounds which contain almost all vertices across our test sets (Fig. 6). At the root level, almost every vertex is contained, with at most 0.1% protruding from the nodes. Furthermore, we find that almost all vertices protruding from their respective bounding spheres are very near the surface of the predicted bounding sphere. As we consider nodes deeper in the trees, we see that at a depth of 7, only around 1% of vertices protrude from their spheres. Various predicted tree depths are shown on a test set deformation in Figs. 9 and 11.

In Fig. 7, we demonstrate that across our test set, we learned the correct radii within at most 1–2%, independent of tree depth. Notably, this percentage error is very tight as compared to a BD-Tree, which often has radial error of over 10–20% across our test sets.

In Fig. 10, we visually compare the results of our method against the BD-Tree method. We find that, even when deformations are small, the BD-Tree method is very conservative with it's bounds, while our method produces bounds much closer to the ground truth.

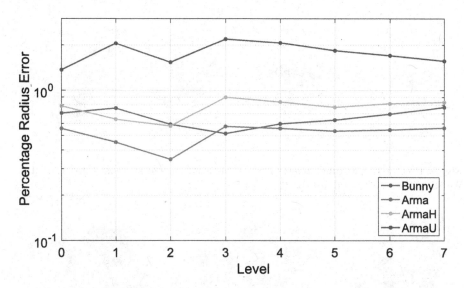

Fig. 7. The percentage error of the sphere radius, measured with respect to the exact spheres (*i.e.*, $10^{-1} = 0.1\%$).

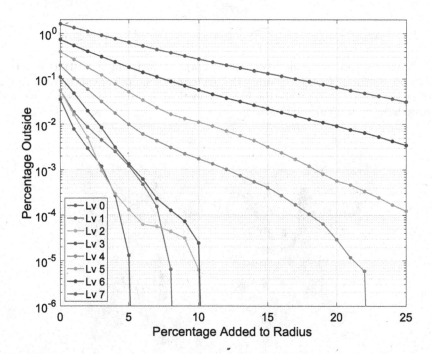

Fig. 8. The percentage of vertices of ARMA that fall outside their predicted node after adding a percentage offset to the predicted radius. The errors fall exponentially.

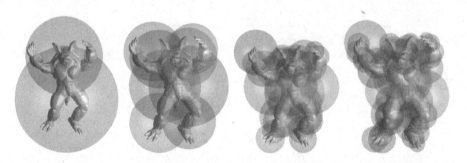

Fig. 9. Predicted bounding spheres for the ARMAHD mesh, at a test set deformation. We show the results of depths 1, 3, 5 and 7.

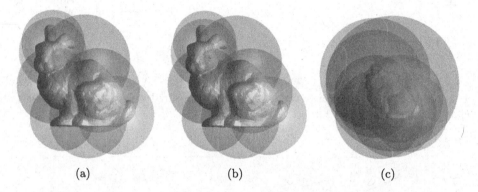

(a)   (b)   (c)

Fig. 10. We compare the results of our method (a) and the traditional BD-Tree method (c) against the ground truth (b) using 128 modes of deformation. Our method closely matches the ground truth, while the BD-Tree method is vastly over conservative with its bounds, even at a mild deformation.

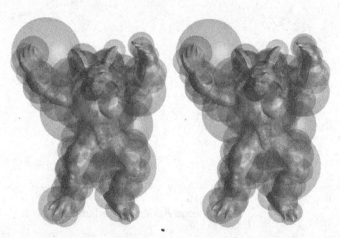

Fig. 11. The predicted bounding spheres at a depth of 7 for a test set deformation of the ARMAUHD model (left), compared to the ground truth (right).

Figure 8 shows how the percentage of vertices of ARMA outside the spheres decreases as we increase the radii by adding a percentage margin. We vary the added percentage from 1% to 25%. The errors for top-level spheres quickly reach zero, whereas the errors for lower-level spheres go down more slowly. However, we note that even for Level 7 spheres, the decrease is exponential.

5   Conclusion and Future Work

We presented NBD-Tree, a neural network-based approach for collision culling of full-space deformable objects. NBD-Trees allows for interactive collision culling for large-scale, full-space deformable objects. It is ideal for collision detection with a highly detailed mesh, possibly driven by lower resolution physics. The performance depends not on the discretization of the mesh but on the number of modes and clusters. Like the classic BD-Tree algorithm, our NBD-Tree algorithm uses low-dimensional modes, but our NBD-Tree algorithm is not limited to reduced-space deformations because we use a neural network to learn the corrections needed to update the spheres to enclose full-space deformations that are in the training set. We compute these low-dimensional modes efficiently and approximately, without going through every vertex of the mesh, relying on the network to overcome the potential errors stemming from these approximations.

Like other learning-based methods, our approach does not work well for extrapolating outside the training set. For extreme deformations that were not seen before, our spheres will inevitably produce more and more false negatives. We also need to train the networks for every new object to be simulated. It would be interesting to see if different discretizations of the same object (*e.g.,* ARMA, ARMAHD, and ARMAUHD) can share the same trained network. This is likely to be the case if the modes are compatible, such as linear modal modes.

If there are many deep inter-penetrations, a standard sphere tree that naively updates its spheres can eventually become cheaper, since their sphere checks are extremely fast. Conversely, our method is extremely efficient if there are few collisions with deep inter-penetrations. Our method is fast at computing the code for the network, but compared to the standard sphere tree, actually carrying out sphere-sphere collisions is more expensive because an MLP needs to be evaluated to compute the center and the radius. Our current implementation based on Eigen can be improved significantly to speed up this process, for example, by using the GPU.

References

1. Alldieck, T., Xu, H., Sminchisescu, C.: imGHUM: implicit generative models of 3D human shape and articulated pose. In: Proceedings of the IEEE/CVF International Conference on Computer Vision, pp. 5461–5470 (2021)
2. Chen, Z., Zhang, H.: Learning implicit fields for generative shape modeling. In: Proceedings of the IEEE/CVF Conference on Computer Vision and Pattern Recognition, pp. 5939–5948 (2019)

3. Deng, B., et al.: NASA neural articulated shape approximation. In: Vedaldi, A., Bischof, H., Brox, T., Frahm, J.-M. (eds.) ECCV 2020. LNCS, vol. 12352, pp. 612–628. Springer, Cham (2020). https://doi.org/10.1007/978-3-030-58571-6_36

4. Ericson, C.: Real-Time Collision Detection. CRC Press, Boca Raton (2004)

5. Faure, F., Allard, J., Falipou, F., Barbier, S.: Image-based collision detection and response between arbitrary volumetric objects. In: ACM Siggraph/Eurographics Symposium on Computer Animation, pp. 155–162. Eurographics Association (2008)

6. Fisher, S., Lin, M.C.: Deformed distance fields for simulation of non-penetrating flexible bodies. In: Magnenat-Thalmann, N., Thalmann, D. (eds.) Computer Animation and Simulation 2001, pp. 99–111. Springer, Heidelberg (2001). https://doi.org/10.1007/978-3-7091-6240-8_10

7. Guibas, L., Nguyen, A., Russel, D., Zhang, L.: Collision detection for deforming necklaces. In: Proceedings of the Eighteenth Annual Symposium on Computational Geometry, SCG 2002, pp. 33–42 (2002). ISBN 1581135041

8. Haddadin, S., De Luca, A., Albu-Schäffer, A.: Robot collisions: a survey on detection, isolation, and identification. IEEE Trans. Rob. **33**(6), 1292–1312 (2017)

9. James, D.L., Pai, D.K.: BD-Tree: output-sensitive collision detection for reduced deformable models. ACM Trans. Graph. **23**(3), 393–398 (2004). ISSN 0730-0301

10. Kingma, D.P., Ba, J.: Adam: A method for stochastic optimization. In: Bengio, Y., LeCun, Y. (eds.) 3rd International Conference on Learning Representations, ICLR 2015, San Diego, CA, USA, 7–9 May 2015, Conference Track Proceedings (2015)

11. Lin, M., Gottschalk, S.: Collision detection between geometric models: a survey. In: Proceedings of IMA Conference on Mathematics of Surfaces, vol. 1, pp. 602–608. Citeseer (1998)

12. Macklin, M., Erleben, K., Müller, M., Chentanez, N., Jeschke, S., Corse, Z.: Local optimization for robust signed distance field collision. Proc. ACM Comput. Graph. Interact. Techn. **3**(1), 1–17 (2020)

13. McAdams, A., Zhu, Y., Selle, A., Empey, M., Tamstorf, R., Teran, J., Sifakis, E.: Efficient elasticity for character skinning with contact and collisions. ACM Trans. Graph. **30**(4) (2011). ISSN 0730-0301

14. Mescheder, L., Oechsle, M., Niemeyer, M., Nowozin, S., Geiger, A.: Occupancy networks: learning 3D reconstruction in function space. In: Proceedings of the IEEE/CVF Conference on Computer Vision and Pattern Recognition, pp. 4460–4470 (2019)

15. Müller, M., Heidelberger, B., Teschner, M., Gross, M.: Meshless deformations based on shape matching. ACM Trans. Graph. **24**(3), 471–478 (2005). ISSN 0730-0301

16. Park, J.J., Florence, P., Straub, J., Newcombe, R., Lovegrove, S.: DeepSDF: learning continuous signed distance functions for shape representation. In: Proceedings of the IEEE/CVF Conference on Computer Vision and Pattern Recognition, pp. 165–174 (2019)

17. Paszke, A., et al.: Pytorch: an imperative style, high-performance deep learning library. In: Wallach, H., Larochelle, H., Beygelzimer, A., d'Alché-Buc, F., Fox, E., Garnett, R. (eds.) Advances in Neural Information Processing Systems, vol. 32, pp. 8024–8035. Curran Associates, Inc. (2019)

18. Pentland, A., Williams, J.: Good vibrations: modal dynamics for graphics and animation, vol. 23, pp. 207–214. ACM, New York (1989). ISSN 0097-8930

19. Santesteban, I., Otaduy, M.A., Casas, D.: SNUG: self-supervised neural dynamic garments. arXiv preprint arXiv:2204.02219 (2022)

20. Seyb, D., Jacobson, A., Nowrouzezahrai, D., Jarosz, W.: Non-linear sphere tracing for rendering deformed signed distance fields. ACM Trans. Graph. **38**(6) (2019). ISSN 0730-0301

21. Sifakis, E., Barbic, J.: FEM simulation of 3D deformable solids: a practitioner's guide to theory, discretization and model reduction. In: ACM SIGGRAPH 2012 Courses, SIGGRAPH 2012. Association for Computing Machinery, New York (2012). ISBN 9781450316781

22. Teschner, M., et al.: Collision detection for deformable objects. In: Computer Graphics Forum, vol. 24, pp. 61–81. Wiley Online Library (2005)

23. Wang, B., Faure, F., Pai, D.K.: Adaptive image-based intersection volume. ACM Trans. Graph. **31**(4) (2012). ISSN 0730-0301

24. Xie, Y., et al.: Neural fields in visual computing and beyond. arXiv preprint arXiv:2111.11426 (2021)

Online Linear Regression Based on Weighted Average

Mohammad Abu-Shaira$^{(\boxtimes)}$ and Greg Speegle

Baylor University, Waco, TX 76798, USA
{mohammad_abu-shaira1,greg_speegle}@baylor.edu

Abstract. Machine Learning requires a large amount of training data in order to build accurate models. Sometimes the data arrives over time, requiring significant storage space and recalculating the model to account for the new data. On-line learning addresses these issues by incrementally modifying the model as data is encountered, and then discarding the data. In this study we introduce a new online linear regression approach. Our approach combines newly arriving data with a previously existing model to create a new model. The introduced model, named *OLR-WA* (OnLine Regression with Weighted Average) uses user-defined weights to provide flexibility in the face of changing data to bias the results in favor of old or new data. We have conducted 2-D and 3-D experiments comparing OLR-WA to a static model using the entire data set. The results show that for consistent data, OLR-WA and the static batch model perform similarly and for varying data, the user can set the OLR-WA to adapt more quickly or to resist change.

Keywords: Online Machine Learning · Weighted Average · Linear Regression · Online Linear Regression · Pseudo-Inverse · Coefficient of Determination (R-squared)

1 Introduction

In Machine Learning, the conventional batch approach operates under the assumption that all data is accessible for every computation. This allows many benefits such as repeatedly accessing the data and cross-validation by leaving portions of the data out. Furthermore, the batch learning approach assumes [8]:

1. the whole training set can be accessed to adjust the model;
2. there are no time restrictions, meaning we have enough time to wait until the model is completely trained;
3. the data distribution does not change; it is typically assumed to be independently and identically distributed (iid). After the model is calibrated, it can produce accurate results without the need for further adjustments.

However, these assumptions limit the applicability of the batch approach [8]. For example consider the scenario of a machine learning model that is trained

© The Author(s), under exclusive license to Springer Nature Switzerland AG 2024
H. Han and E. Baker (Eds.): SDSC 2023, CCIS 2113, pp. 88–108, 2024.
https://doi.org/10.1007/978-3-031-61816-1_7

to predict stock prices. The model is initially trained on historical stock market data and is used to make predictions about future stock prices. However, as time passes, the stock market changes. For example, the economy can go through a recession or a period of high inflation, new companies can go public, and old companies can go bankrupt. These changes in the stock market mean that the initial training set used to train the model is no longer valid. In order to continue making accurate predictions, the model must be updated to adapt it to the new conditions. Another scenario is in streaming environments where predictions are required at any given moment during execution. In such cases, the batch model must be recreated each time a prediction is required [8]. For example, consider the scenario of a traffic management system in a smart city. In this case, the system needs to provide real-time predictions for traffic flow and congestion levels at different locations. As traffic conditions can change rapidly due to accidents, road closures, or unexpected events, the system must recreate its prediction model each time a new prediction is required. By continuously updating the model with the latest traffic data, the system can offer accurate and up-to-date information to drivers, allowing them to choose the most efficient routes and alleviate congestion in real-time.

By considering these restrictions, we come to the realization that the applicability of machine learning is greatly constrained. Many significant applications of learning methods in the past 50 years would have been impossible to solve without relaxing these restrictions. On the other hand, online learning assumes: [8]:

1. only a portion of the data is available at any one time
2. the response should be timely
3. the data distribution can change over time

This study introduces a novel online linear regression model *Online Regression with Weighted Average (OLR-WA)* that is based on the weighted average of a base model which represents the already seen data and an incremental model which represents the new data. OLR-WA eliminates the challenge of storage requirements for large amounts of data while providing an effective solution for large-scale problems. Additionally, it does not require any assumptions about the distribution of data and instead can work in an adversarial scenario, making it adaptable to a wide range of situations where the data may not be independently and identically distributed.

This paper makes two significant contributions. First, the OLR-WA model performs comparably to a batch model over data consistent with the batch model expectations. Second, the OLR-WA model provides flexibility not found in the batch model and many other online models.

2   Related Work

In this section, we review some of the work related to online learning and to linear regression.

2.1   Stochastic Gradient Descent (SGD)

Gradient descent is a commonly used optimization algorithm for training machine learning models. It is based on the idea of iteratively adjusting the model's parameters in the direction of the negative gradient of the loss function, thereby minimizing the loss. Stochastic gradient descent (SGD) is a variation of gradient descent that uses a random sample of the data, rather than the full dataset, to compute the gradient at each iteration. This makes SGD more computationally efficient. SGD is one of the most widely used techniques for online optimization in machine learning [2]. Incremental algorithms, like Stochastic Gradient Descent (SGD) have been found to be more effective on large data sets than batch algorithms, and are widely used [9]. Mini-batch gradient descent combines the two approaches by performing an update for every mini-batch of n training examples [16].

SGD and mini-batch SGD are well suited to online learning and widely used in the industry [2]. While SGD and mini-batch SGD can be used for online learning, it works under the assumption that all the observed data up to the present moment is consistently accessible. They process the data one sample or a small batch of samples at a time, updating the model parameters after each sample or batch. In addition to that, these methods suffer from different limitations like [16] the sensitivity to the choice of the learning rate, which affects the convergence speed. A learning rate that is too small will result in slow convergence, while a learning rate that is too large may cause the algorithm to oscillate or not converge at all. Additionally, SGD might get stuck in a local minimum, which can result in poor predictions.

2.2   Linear Regression

Linear Regression [12, 20] is one of the most common and comprehensive statistical and machine learning algorithms. It is used to find the relationship between one or many independent variables and one dependent variable. It is a mathematical approach used to perform predictive analysis and can be used to determine causal relationships in some cases.

Regression may either be simple or multiple regression. Simple linear regression studies the relationship between two continuous (quantitative) variables. One variable, denoted 'x', is regarded as the predictor, explanatory, or independent variable and the other variable, denoted 'y', is regarded as the response, outcome, or dependent variable [12]. The model equation is represented by $y = \beta_0 + \beta_1 x + \epsilon$ Multivariate linear regression (MLR) is used to predict the result of an answer variable using a number of explanatory variables. The basic model for MLR is $y = \beta_0 + \beta_1 x_1 + \cdots + \beta_m x_m + \epsilon$. The formula to determine the formula matrix (usually called pseudo-inverse) is [12]

$$\hat{\beta} = (X^T X)^{-1} X^T \mathbf{y}$$

where $\beta = \begin{bmatrix} \beta_0 \\ \beta_1 \\ \vdots \\ \beta_m \end{bmatrix}$, $X = \begin{bmatrix} 1 & x_{11} & x_{12} & \ldots & x_{1m} \\ 1 & x_{21} & x_{22} & \ldots & x_{2m} \\ \vdots & \vdots & \vdots & \vdots & \vdots \\ 1 & x_{n1} & x_{n2} & \ldots & x_{nm} \end{bmatrix}$, $y = \begin{bmatrix} y1 \\ y2 \\ \vdots \\ y_n \end{bmatrix}$

2.3 On-Line Linear Regression Models

In general, on-line learning models follow the framework in Algorithm 1, which aims to minimize the total loss incurred.

Algorithm 1. General Online Learning Framework [17]

1: Initialize $w_1 = 0$
2: **for** each round t=1,...,T: **do**
3:      Get training instance x_t
4:      Predict label $\hat{y}_t \in \mathbb{R}$
5:      Get true label $y_t \in \mathbb{R}$
6:      Incur loss $l(\hat{y}_t, y_t, x_t)$
7:      Update w_t
8: **end for**

Widrow-Hoff. One of the on-line linear regression models is the Widrow-Hoff algorithm, also known as Least-Mean-Square (LMS) Algorithm. LMS combines stochastic gradient descent techniques with a linear regression objective function by considering data one point at a time. For each data point, the algorithm makes successive corrections to the weight vector in the direction of the negative of the gradient vector. This eventually leads to the minimum mean square error. The LMS update rule is:

$$w_{t+1} = w_t - \alpha(w_t.x_t - y_t)x_t \tag{1}$$

where α is the learning rate, and w_t is the weight vector of the current iteration. This update rule has been derived so we can stay close to w_t, since w_t embodies all of the training examples we have seen so far. In our work, we call this confidence bias, and can be modeled within OLR-WA [8,13].

The Widrow-Hoff algorithm is commonly used adaptive algorithm due to its simplicity and good performance. However, the value of the learning rate parameter must be chosen carefully to ensure the algorithm converges. As an iterative algorithm, it can adapt to a rapidly changing data environments, but its convergence speed may be slower than other algorithms. Additionally, the LMS algorithm has a fixed step size for each iteration and may not perform well in situations where the input signal's statistics are not well understood or bursty [8].

Online Support Vector Regression. Support Vector Regression (SVR) was introduced in the 1990s by Vadimir Vapnik and his colleagues (Drucker, Cortes, & Vapnik, 1996) [7] while working at AT&T Bell Labs. The detailed exploration of SVR can be found in Vapnik's book (Vapnik, 1999) [18]. Vapnik's SVR model distinguished itself from standard regression models by fitting a tube, commonly known as the ϵ-Insensitive Tube, instead of a line. This tube, with a width denoted as $\epsilon > 0$, defines two sets of points: those falling inside the tube, which are considered ϵ-close to the predicted function and are not penalized, and those falling outside the tube, which are penalized based on their distance from the predicted function. This penalty mechanism bears similarity to the penalization used by Support Vector Machines (SVMs) in classification tasks. In addition to the tube-based approach, SVR incorporates a kernel function. This kernel function allows SVR to capture nonlinear relationships between the input features and the target variable. It provides the flexibility to choose appropriate transformations for diverse types of data and problem domains.

The online variant of SVR employs stochastic gradient descent (SGD), which applies the concept of updating the dual variables incrementally based on the deviation between the predicted and target values. The dual variables, typically represented by α, are optimization variables associated with each data point. These variables measure the importance or weights assigned to each data point in the training set. The values of the dual variables determine the influence of each data point on the final decision function. Data points with non-zero dual variables, referred to as support vectors, play a significant role in defining the decision boundary or regression surface.

Our approach fundamentally diverges from Online Support Vector Regression (Online SVR) in several key aspects. Firstly, Online SVR randomly selects one data point at a time from the entire pool of previously observed data using stochastic gradient descent. In contrast, our approach requires a minimum number of data points, forming a mini-batch, to construct a model. Moreover, our approach encompasses the ability to forget previously seen data points while preserving their associated metadata in the form of a weighted average generated model. This distinguishing feature of our approach proves advantageous, particularly in adversarial scenarios, as the weights can be tailored to favor user-specified criteria. By leveraging this weighted average model, our approach demonstrates flexibility and adaptability in such scenarios.

Recursive Least-Squares (RLS) Algorithm. The Recursive Least-Squares (RLS) algorithm is a type of adaptive filter algorithm that is used to estimate the parameters of an online linear regression model. It is a recursive algorithm that uses a least-squares criterion to minimize the error between the desired output and the estimated output of the system. In the context of online linear regression, the RLS algorithm is used to estimate the coefficients of the linear model. The algorithm starts with an initial estimate of the coefficients and updates them in real-time based on new data points. The algorithm uses a recursive update rule to adjust the coefficients based on the current data point and the previous estimates. The update rule is based on the gradient descent optimization method, which aims to minimize the mean square error between the desired output and the estimated output. The RLS algorithm uses a forgetting factor, which is a scalar value between 0 and 1, to balance the trade-off between the importance of the current data point and the importance of previous data points. A higher forgetting factor value gives more weight to the current data point, while a lower forgetting factor value gives more weight to previous data points. Similar weights are used in OLR-WA. In summary, the RLS algorithm is an efficient and robust algorithm that can estimate the parameters of online linear regression models in real-time. It has fast convergence and good performance in terms of stability and robustness. However, it can be computationally expensive when the number of input variables is large [8].

Our approach is fundamentally different from either LMS or RLS in that OLR-WA computes an incremental model based upon a collection of inputs and then integrates the incremental model into an existing base model representing all of the previous data. This allows model integration instead of data integration, which means larger bursts of data points can be handled and we can emphasize either the previous model, the incremental model or neither. While we intend to compare both performance and accuracy of OLR-WA with other techniques in the future, in this paper we validate OLR-WA by comparing it to linear regression with all data available.

Online Ridge Regression. Ridge regression is a regression technique that addresses the issue of overfitting in linear regression models by introducing a regularization term. Overfitting occurs when a model fits the training data too closely, resulting in poor performance on new, unseen data. Ridge regression adds a penalty term to the loss function during training to constrain the model's coefficients, thus reducing overfitting. [19] The cost function in ridge regression consists of two parts: the residual sum of squares (RSS) term $(y - Xw)^T(y - Xw)$ that measures the discrepancy between the predicted and actual values, and the regularization term $\lambda w^T w$ that penalizes large coefficient values to prevent overfitting. The regularization term helps in controlling the complexity of the model and reducing the impact of irrelevant features. [13] By applying stochastic gradient technique to ridge regression, we can derive a similar algorithm. In each round, the weight vector is updated with a quantity based on the prediction error $(y_t - X_t w_t)$. The main idea behind Online Ridge Regression is to update the model's parameters in an incremental fashion while incorporating regularization. This is achieved by adapting the standard ridge regression algorithm to handle streaming data. The difference between regular ridge regression and online ridge regression lies in the way the data is processed and updated. In online ridge regression, the data is processed sequentially, updating the coefficient vector w after each observation. Here's the formula for online ridge regression:

$$J(w) = (y - Xw)^T(y - Xw) + \lambda w^T w \tag{2}$$

where $J(w)$ represents the cost function, y represents the vector of observed or target values, X represents the matrix of predictor variables or features, w represents the coefficient vector or parameter vector, which contains the regression coefficients for each feature, λ represents the regularization parameter, which controls the trade-off between fitting the training data and preventing overfitting, $(y - Xw)^T (y - Xw)$ represents the squared residual term, which measures the difference between the observed values y and the predicted values obtained by multiplying the predictor variables X with the coefficient vector w, and finally $\lambda w^T w$ represents the regularization term, which penalizes the magnitude of the coefficient vector w to prevent overfitting.

Our methodology takes a fundamentally different approach compared to Online Ridge Regression across several critical aspects. Firstly, whereas Online Ridge Regression utilizes stochastic gradient descent to randomly select individual data points from the entire pool of previously observed data, our approach requires a minimum number of data points to form a mini-batch for constructing the model. Furthermore, our method allows for the forgetting of previously encountered data points while retaining their associated metadata through the generation of a weighted average model. This distinctive feature of our approach offers clear advantages, particularly in adversarial scenarios, as the weights can be tailored to prioritize specific user-defined criteria. By leveraging this weighted average model, our approach demonstrates remarkable flexibility and adaptability in such challenging situations.

The Online Passive-Aggressive (PA) Algorithms. The Passive-Aggressive algorithms is a family of algorithms for online learning. It is often used for classification tasks but can also be applied to regression. The algorithm updates the model's parameters in a way that minimizes the loss using the below update rule

$$w_{t+1} = \arg \min_{w \in \mathbb{R}^n} \frac{1}{2} \|w - w_t\|^2 \qquad s.t \quad l(w; (x_t, y_t)) = 0 \tag{3}$$

while remaining "passive" whenever the loss is zero, that is $w_{t+1} = w_t$ or "aggressive" in which those rounds the loss is positive, then the algorithm aggressively forces w_{t+1} to satisfy the constraint $l(w_{t+1}; (x_t, y_t)) = 0$. In addition to that, the algorithm has two other variations $PA - I$, and $PA - I$, which adds the terms $C\xi$, and $C\xi^2$ respectively. C the "aggressiveness parameter" is a positive parameter which controls the influence of the slack term ξ on the objective function. Larger values of C imply a more aggressive update step. In other words, the parameter C represents the regularization parameter, and denotes the penalization the model will make on an incorrect prediction.

By repeating the training process for each training example, the Online Passive-Aggressive algorithm adapts its parameters to minimize the loss while considering the aggressiveness of the updates. The aggressiveness of the updates allows the algorithm to quickly adapt to new patterns in the data. Overall, the Online Passive-Aggressive algorithm is a useful tool for online learning tasks, including online linear regression. It can adapt to changing data streams and make updates to the model's parameters based on the aggressiveness determined by the training examples encountered [6].

The Passive Aggressive Online Algorithm is an efficient approach for learning on the fly in scenarios involving a continuous stream of data with labeled documents arriving sequentially. An illustrative use case is monitoring the entire Twitter feed 24/7, where each individual tweet holds valuable insights for prediction purposes. Due to the impossibility of storing or retaining all tweets in memory, this algorithm optimally processes each tweet by promptly learning from it and subsequently discarding it. The PA Online Algorithm, similar in some respects to the OLR-WA, possesses a mechanism for discarding data points. However, OLR-WA distinguishes itself from PA algorithm by incorporating weights that favor either the base or incremental model. However, PA algorithm exhibits a default behavior that leans towards favoring new incoming data points.

3   OLR-WA Methodology

The OLR-WA algorithm creates a new linear regression model for each data sample. The sample model and the existing model are merged to form a new linear regression model which can be used to make predictions until the next sample arrives. We compare OLR-WA to a batch linear regression approach which waits until all of the data arrives to build a model. In addition to being able to make predicitions sooner, the final result of OLR-WA has similar results as to the batch model.

3.1   Data Sets

In this study we use relatively small synthetic and real world data sets. Each synthetic data set is drawn from either a two dimensional data distribution or a three-dimensional data distribution. For our experiments, we consider three types of distributions. The first is where all of the data is from the same linear distribution with a low noise factor as variance. The second is also a linear distribution, but the variance is higher. The third data set is a combination of two data sets, each with different liner distributions. This is an adversarial situation representing a change in the data distribution. Each experiment is represented by the number of data points "N", the variance "Var", the correlation "Cor" if it is positive, or negative, and the step size between data points "Step". the following figures show some distribution samples of the dataset with 2-D settings.

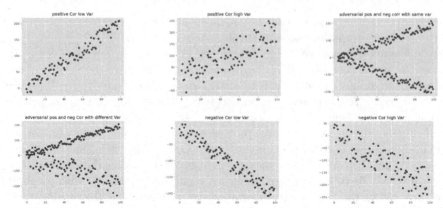

In addition to that, for the purpose of validating our model with real world data sets, we used two public data sets explained in Sect. 4.4.

3.2 Method

Our approach includes two parts: a base model and an incremental model. The base model is the initial set of data points used as a starting point for linear regression. There is no specific limit to the number of data points, the minimum number necessary to create a model (e.g. 2 points for a 2-D model, 3 points for a 3-D model). The incremental model is a linear regression model that is created with each new set of data points. Again, there is no limit to the number of points to create a model, but we use the same guidelines as for the base model, (e.g. 2 points for a 2-D model, 3 points for a 3-D model). In our experiments, we allocate 10% of the data points for the base model, while the remaining points are added incrementally at a rate of 10 points per increment.

Once both models are created, we calculate their weighted average. For 2D, this means averaging two lines, for 3D is averaging two planes and in higher dimensions, we must average hyperplanes. The weighted average is computed by assigning user-defined weights to each model. This allows the user to modify the results based upon their knowledge of past and future data. If the incremental model is given higher weight, the model adapts to changing data more quickly. If the base model is given higher weight, the model is more resistant to transient changes. By default, the weights are equal, where both w-base and w-inc are assigned a fixed static number. Later in the paper, we will elaborate on various options for weight settings.

Although the techniques for computing the weighted average of lines, planes, or spaces may differ, the basic equation used remains consistent.

$$\textbf{V-Avg} = (\textbf{w-base} \cdot \textbf{v-base} + \textbf{w-inc} \cdot \textbf{v-inc})/(\textbf{w-base} + \textbf{w-inc}) \qquad (4)$$

where **w-base** represents the weight we assign to the base model, **w-inc** represents the weight we assign to the incremental model, **v-base** is the vector of the base model and **v-inc** is the vector of the incremental model. The weights here are scalar values, but the models will differ based on the number of dimensions.

For example, in the 2D case, the base model is a line, so **v-base** will be computed using the equation

$$V = \langle x2 - x1, y2 - y1 \rangle \tag{5}$$

where (x1, y1) and (x2, y2) are the coordinates of the tail and head of the line respectively. Similarly, in 3D, if the normal vector of the plane equation $13x + 3y - 6z = 15$ then $V = \langle 13, 3, -6 \rangle$ The same applies for **v-inc**. **V-Avg** is the computed average vector, which we'll use later on with intersection point of the two models to construct the average line, plane, or space.

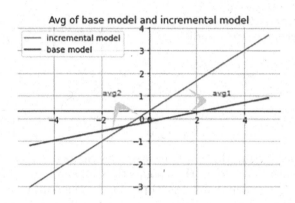

Fig. 1. Two computed averages.

We can compute two weighted average vectors from the base model and the incremental model, Fig. 1 shows the idea in two-dimensional plane in which the linear regression models for the base and the increment are represented by lines. The formula generates two averages, labeled avg1 and avg2 in Fig. 1. OLR-WA picks one of these two average vectors by detecting which one fits the data better. Since only the incremental data is available for evaluation, OLR-WA generates previous data from the base model. Using the incremental data and the generated data, the model selects the best fit. For example, in Figure 1 avg1 should be retained as the new base model, since it will have less mean square error. Algorithm 2 shows the steps of OLR-WA.

We explore different techniques for assigning weights to data points, each with its own feasibility, correctness, and significance. Let's discuss each assignment technique along with short examples:

a) Time-based: [4], this technique assigns weights to data points based on their age or recency. The weight of a data point decreases as it becomes older. This method acknowledges that recent data points are more likely to be relevant to the current situation and gives them more importance. For instance, in a stock market prediction system, recent stock prices might carry more weight in determining the future trend compared to older prices.

Algorithm 2. Online Linear Regression with Weighted Average

1: base-regression = pseudo-inverse(base-X,base-y)
2: **for** $t \leftarrow 1$ to T **do**
3:     inc-regression = pseudo-inverse(inc-X,in-y)
4:     **v-avg1** = (**w-base** · **v-base** + **w-inc** · **v-inc**)/(**w-base** + **w-inc**)
5:     **v-avg2** = (-1 · **w-base** · **v-base** + **w-inc** · **v-inc**)/(**w-base** + **w-inc**)
6:     intersection-point = get-intersection-point(base-regression, inc-regression)
7:     space-coeff-1 = define-new-space(**v-avg1**, intersection-point)
8:     space-coeff-2 = define-new-space(**v-avg2**, intersection-point)
9:     err-v1= MSE(space-coeff-1)
10:     err-v2= MSE(space-coeff-2)
11:     **if** err-v1 < err-v2 **then**
12:         coefficients ← space-coeff-1
13:     **else**
14:         coefficients ← space-coeff-2
15:     **end if**
16: **end for**
17: **return** coefficients

---

b) Confidence-based: [15], this technique assigns weights based on the confidence or accuracy of the data point. Data points that are known to be more accurate or reliable are given higher weights. For example, in a sentiment analysis task, if certain labeled data points have been verified by experts or trusted sources, those points could be assigned higher weights due to their higher confidence level.

c) Fixed-based: [21], our model here offers the flexibility for two variations. The first variation assigns fixed equal weights to all data points, treating them equally in terms of importance. This implies that every data point contributes equally to the model. For instance, in a weather prediction model, each recorded temperature measurement might have the same weight. In this variation, the weight of the base model, w-base, is updated after each iteration by adding the weight of the incremental model, w-inc. This update is necessary because the current generated model incorporates both the base data and the incremental data, reflecting their combined influence. The second variation of the fixed-based technique assigns fixed equal weights to both the base model and the incremental model. This signifies that the existing state of the base model holds equal weight to the new model generated based on the incremental data. This ensures a balanced integration of the existing and new information. For example, in a machine translation system, both the previously trained model and the additional training data have equal importance in generating accurate translations.

One of the most common techniques is called the Decay Factor, which is a value used to reduce the weight of older data points in an online learning algorithm. The decay factor is multiplied by the weight of each data point at each iteration, so older data points have a lower weight than newer data points.

This approach gives more importance to recent data points, as they are more likely to be relevant to the current situation [11]. The decay factor is considered a time-based weighting scheme. It is important to note that the choice of weighting scheme depends on the specific problem and dataset at hand, and it's important to experiment with different weighting schemes to find the one that works best for the specific use case. In the experiments section, we will demonstrate some techniques for adjusting weights to achieve the desired outcomes.

3.3   Time Complexity

The algorithm presented can be applied to any form of linear regression, but using the pseudo-inverse is a particularly common choice as it is polynomial time. We will be focusing on this approach in our analysis of the algorithm. The algorithm executes one pseudo inverse linear regression for the base model, and T pseudo inverse linear regression for incremental data, the pseudo inverse linear regression equation as we stated earlier is $\hat{\beta} = (X^T X)^{-1} X^T \mathbf{y}$.

Given X to be a M by N matrix, where M is the number of samples and N the number of features. The matrix multiplications each require $O(N^2 M)$, while multiplying a matrix by a vector is $O(NM)$. Computing the inverse requires $O(N^3)$ in order to compute the LU or (Cholesky) factorization. Asymptotically, $O(N^2 M)$ dominates $O(NM)$ so we can ignore that calculation. Since we're using the normal equation we will assume that M > N, otherwise the matrix $X^T X$ would be singular (and hence non-invertible), which means that $O(N^2 M)$ asymptotically dominates $O(N^3)$. Therefore, the total complexity for the pseudo inverse is $O(N^2 M)$.

Our proposed algorithm uses a linear regression technique, specifically the pseudo-inverse linear regression, to process the data incrementally. This allows for the processing of smaller batches of data at a time, resulting in a lower overall data size. Compared to the traditional batch version of the pseudo-inverse linear regression, which has a time complexity of $O(N^2 M)$, our online model's time complexity is likely to be $O(KN^2(M/K))$, where K is the number of iterations. Thus the total time complexity of OLR-WA is about the same as for the batch model running all at once.

3.4   Evaluation Metric

The coefficient of determination, "usually denoted by $R2$ or $r2$, is the proportion of variation of one variable (objective variable or response) explained by other variables (explanatory variables) in regression,ch7kasuya2019use". This is a widely used measure of the strength of the relationship in regression. It describes how well the model fits the data. An r^2 close to 1 implies an almost perfect relationship between the model and the data [14]. This coefficient is defined as [10]

$$r^2 = 1 - \frac{\sum_{i=1}^{n}(y_i - \hat{y}_i)^2}{\sum_{i=1}^{n}(y_i - \overline{y}_i)^2} = 1 - \frac{SE\,\hat{\mathbf{y}}}{SE\,\overline{\mathbf{y}}} \tag{6}$$

where ŷ denotes the value of the objective variable (y) predicted by regression for the ith data point. The second term of this expression is the residual sum of squares divided by the sum of squares of y.

It is highly recommended to use the coefficient of determination as the standard metric for evaluating regression analyses in any scientific field because it is more informative and accurate than SMAPE, and does not have the interpretability limitations of other metrics such as MSE, RMSE, MAE, and MAPE [3]. The coefficient of determination is used as the evaluation metric for our model.

4   Discussion

In this research, we carried out experiments utilizing both 2-dimensional and 3-dimensional datasets. We employed the versatility and management capabilities of our dataset generator to perform multiple experiments. We will now examine the model's behavior and implications.

4.1   2-Dimensional

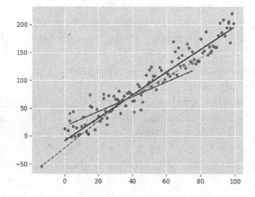

Fig. 2. Two average lines (Color figure online)

Figure 2 provides a comprehensive illustration of a 2D scenario, the green points represent the base model points, which are used for visualization purposes only, but in reality, we don't retain them, the only thing we maintain about the those base points is the model itself which is represented here by a the green line. The blue points are the new coming (incremental) points, the blue line represent the linear regression line for the incremental points. The yellow dashed line ends with a red star on its tail represents first computed average line, and the gray dashed line ends with a red star on its tail represents the second computed linear regression line.

In the 2D model, we define the average lines, the yellow and the gray in Fig. 2 by the two norm vectors resulted from Eq. 4 and a point of intersection between the base and the incremental lines. It is worth nothing that it is highly unlikely for the base and the incremental lines to have exactly the same slope, making them parallel, and thus having no intersection point. In this scenario, the current algorithm simply ignores the case and updates itself on the next iteration. Although this is a rare occurrence, for more accurate results, there are several solutions which we will consider in the future work.

As illustrated earlier, one of those two lines will be selected and will represent our current model, the other will be discarded selection is based on the minimum Mean Square Error (MSE) of the new coming data and some sampled data from the base model.

4.2   3 Dimensional

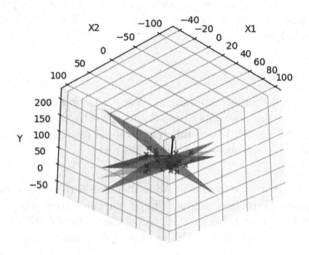

Fig. 3. Two average planes.

Figure 3 provides a comprehensive illustration of a 3D scenario, the blue plane represent the base model. The brown plane represents the incremental model. The orange plane represents first computed average plane, and the green plane represents the second computed plane. The 3D differs from the 2D with the coefficients. In 2D the coefficients are the slope or m_0, and the y_intercept or (b), while in 3D the coefficients are m_0, m_1, m_2 or (b). In the 3D we define the average planes, the orange and the green in Fig. 3 by the two norm vectors resulted from Eq. 4 and a point of intersection between the base and the incremental plane. In this case, the intersection of two planes is a line, so we can use any point on the line. In general, the intersection point of two planes or spaces can be determined by solving the equation of both the base and incremental models, which will provide a point that is located within both planes or spaces.

4.3   Generalization to Higher Dimensions (N-Dimensional)

While OLR-WA is straightforward in 2-D and 3-D, the same procedure can be applied to higher dimensions. we only need to find a point in the intersection and the direction vector, and using those two elements, we can compute the new hyperplane. Obtaining the point of intersection differs from one dimensional space to another. For example, in 2-D, finding the intersection of two lines requires solving two equations with two variables, which results in an exact one point. However, in 3-D, finding the intersecting line required solving two equations with three variables, which produces an infinite number of solutions which are points representing that line. In higher dimensions, the intersection of N-Dimensional hyperplanes, is a (N-1)-Dimensional hyperplane which will be generated by solving two equations of N variables. This can be done using methods like Gaussian elimination, matrix inversion, or using software packages. The solution will be a point $(x_1, x_2, ..., x_n)$ that satisfies both equations, representing the intersection of the two hyperplanes. Note that there can be multiple solutions, depending on the hyperplanes' orientation and position in the N-dimensional space.

4.4   Experiments

We have conducted several experiments using 2-D and 3-D, we used 10% of the total points as the base model, and a total of 10 points on each iteration for the incremental model. In the following we will provide a summary of these experiments and the outcomes that were obtained.

Experiment 1. In this experiment we generated 5 different datasets of 200 random, positively correlated points. The r^2 for the batch model ranged from 0.9152 (in 3D) to 0.9558 (also in 3D), while the r^2 for OLR-WA ranged from 0.8991 (in 2D) to 0.9517 (in 3D). Tables 1a and 1b show the sample runs of the experiment.

Experiment 2. In this experiment, we generated 5 different datasets of 200 random, positively correlated points. However, this time 100 data points are randomly generated with one variance and the other 100 data points are generated with a different variance. Not surprisingly, neither model was able to model the data as accurately as Experiment 1, with the r^2 for the batch model ranging from 0.8222 (in 3d) to 0.9190 (in 2D) while OLR-WA r^2 ranges from 0.8107 (in 3D) to 0.8904 (in 2D). Tables 2a and 2b shows the sample runs of the experiment.

Experiment 3. The aim of the this experiment is to expose OLR-WA represented by Algorithm 2 into real world datasets and validate its performance, we used two public datasets:

Table 1. Results of 2D Experiments

Experiment	Batch	Online		Experiment	Batch	Online
1	0.9348	0.9331		1	0.8444	0.8340
2	0.9492	0.9349		2	0.8710	0.8323
3	0.9379	0.8991		3	0.8982	0.8869
4	0.9388	0.9291		4	0.9190	0.8904
5	0.9250	0.9134		5	0.8795	0.8695

(a) Batch vs. Online for Positively Correlated Data   (b) Batch Vs. Online for Dataset with Shifting Variance

Table 2. Results of 3D Experiments

Experiment	Batch	Online		Experiment	Batch	Online
1	0.9348	0.9326		1	0.8400	0.8280
2	0.9267	0.9080		2	0.8738	0.8591
3	0.9558	0.9517		3	0.8222	0.8209
4	0.9152	0.9156		4	0.8483	0.8470
5	0.9422	0.9449		5	0.8259	0.8107

(a) Batch vs. Online for Positively Correlated Data   (b) Batch vs. Online for Data with Shifting Variance

1. [1] 1000 companies data set. The dataset includes sample data of 1000 startup companies operating cost and their profit. Well-formatted dataset for building ML regression pipelines. This data set is used to predict profit from R&D spend and Marketing spend
2. [5] Math Student data set. This is a dataset from the UCI datasets repository. This dataset contains the final scores of students at the end of a math programs with several features that might or might not impact the future outcome of these students. Math Student dataset is used to predict secondary school student's performance (final grade) using first period grade, and second period grade.

Tables 3a and 3b show the performance of OLR-WA versus the standard batch model using the 2 aforementioned data sets.

Experiment 4. In this experiment we show an adversarial scenario. As in Experiment 2, we generate 200 random data points. However, in this case, the first 100 points are positively correlated while the next 100 points are negatively correlated. As expected, the batch model does not generate strong r^2 results (and it can be argued that is correct), see Fig. 8. However, in the online approach, the user can supply weights to indicate their preference for the data. For example, older points can be given higher, lower or the same weight as newer points. We

Table 3. Results of Experiments on Real Data Sets.

Experiment	Batch	Online
1	0.8078	0.7420
2	0.8151	0.8001
3	0.9979	0.9968
4	0.6972	0.5949
5	0.9976	0.9417

(a) Batch vs. Online for [1]
1000 Companies Data Det.

Experiment	Batch	Online
1	0.8027	0.7913
2	0.7829	0.7818
3	0.8453	0.8448
4	0.8302	0.8118
5	0.8802	0.8759

(b) Batch vs. Online for [5]
Math Student Data Set

classify the weights as **time-based** in which older points have lower weights, **confidence-based** in which older points have higher weights and **fixed-based** in which the relative weights are established a priori.

Fixed-Based Weights. The default case, known as the fixed-based weights case, presents two options for the user's consideration. Firstly, all points are assumed to be treated equally, resulting in equal weights for each point. Secondly, both models are assigned fixed equal weights throughout the process. We will delve into these two cases in the following sections:

In the first case, assuming equal weights for all points, the base model begins with 40 points, denoted as w-base = 40. Meanwhile, the incremental model processes 10 points during each iteration, denoted as w-inc = 10. After each iteration, the weight of the base model increases as it accumulates the incremental points, given by the equation w-base += w-inc. Consequently, the base model progressively carries a higher weight than the incremental model. As a result, the model's regression plane aligns with the base model, as illustrated in Fig. 4.

In the second case, the user assigns fixed equal weights to both models from the outset. Specifically, the user designates w-base = 1 and w-inc = 2. In this scenario, the regression plane gradually shifts towards the incremental model and aligns with it, as depicted in Fig. 5.

Time-Based Weights. Figure 6 shows the same distribution (but different data points), but this time the incremental model has a weight 20 times greater than the existing model. This represents a scenario in which the model trusts the data will continue with the new distribution. Note that the plane is a good fit for the incremental model.

Confidence-Base Weights. Figure 7 again has the same data distribution, but different data points. In this case, the weight of the existing model is 20 times higher than the weight of the incremental model. This scenario entails placing greater trust in the initial data while assigning minimal weight to the incremental data in order to diminish its impact. Note that the plane is a good fit for the existing model.

Between the three models, the time-base approach is the most dynamic. For example, a 3rd increment with yet another distribution would cause the model to change to match the new data, while the fixed weights would change less and the confidence weights the least.

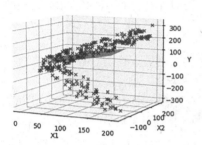

Fig. 4. Fixed-Based Weights - equal points weights

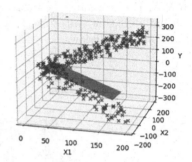

Fig. 5. Fixed-Based Weights - equal models weights

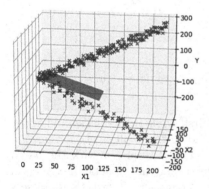

Fig. 6. Time-Based Weights

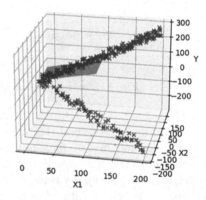

Fig. 7. Confidence-Based Weights

5   Conclusion and Future Work

We have conducted 2D, and 3D experiments using OLR-WA and a batch model. In general, OLR-WA preformed comparable to the batch model. For example, in the 2D case with constant variance, the r^2 values for the batch model varied from 0.9251 to 0.9492 while the r^2 for OLR-WA varied from 0.8991 to 0.9349. Similar results hold for the experiments with shifting variance and the 3D experiments. Furthermore, the time complexity of OLR-WA is on par with the batch version. Additionally, it provides flexibility that is not a standard part of the batch model by allowing dynamic adjustment of weights. In other words, the standard batch model lacks the flexibility required to handle adversarial scenarios like OLR-WA. For instance, as illustrated in Fig. 8, when incremental data has a significant shift that calls for a new model, the r^2 of the standard batch model will be considerably low.

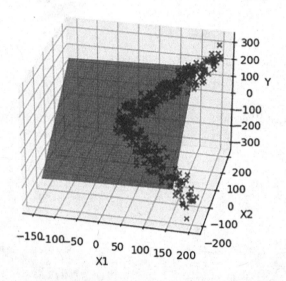

Fig. 8. Standard Batch Model with Adversarial Scenario.

The work can be extended in many ways. First, while the model should adapt to any dimensionality, we have not tested extending the implementation beyond 3D. Second, while the time complexity of the model is on par with the batch model, we have not run performance tests to determine the impact of incremental evaluation, especially with respect to very large data sets. Third, the data sets were designed to find good fits. We need to compare the batch method with OLR-WA on datasets with significantly greater noise (much lower r^2 values). One issue specific to OLR-WA that can be addressed is the case where the two models do not intersect. One option is finding the closest distance between the two lines that can be used instead or possibly increasing the incremental model size to resolve the problem of parallelism.

While OLR-WA shows promise when compared to the batch model, we also need to compare it to other incremental learning techniques, such as LMS and RLS. The comparisons need to consider data distributions, adversarial situations and large data sets (both in terms of the number of features and the number of points).

Finally, OLR-WA has potential for interesting extensions, specifically in the area of weight selection. Currently, weights (w-base, and w-inc) are chosen by the user, but it would be fascinating to give OLR-WA the ability to select weights based on both user preferences and/or observed data. By recording data such as the number of points, variance, and correlation of the base model and each incremental model, as well as the user's preferred weight classes (time, confidence, or fixed), OLR-WA can automatically select weights. For instance, if the user favors time-based weights and the incremental model significantly deviates from the existing model, OLR-WA could automatically increase the weight of w-inc which represents the incremental data weight. Likewise, when presented with adversarial data, OLR-WA could automatically increase the size of the incremental batch or consider many mini-batches to allow more points to accumulate and determine whether the new data fits a different model or represents an outlier, if it is highly likely that the adversarial data is constituting a new different model, and the user favors the weight-time weights, then, OLR-WA automatically will increase w-inc to the favor on the new model, and similarly if the user favors the base model to a certain extent, OLR-WA will automatically increase w-base which represents the base model weight till that point in time. Another example will be computing the incoming tweet or post weight based on the likes or positive/negative comments it receives. Furthermore, introducing a forgetting factor can enhance this capability. When the model detects the introduction of a new model through a series of incremental mini-batches, it can be programmed to forget the old model and begin considering the new model. This approach increases the flexibility of the model and enables it to adapt to changing data distributions more effectively.

References

1. 1000 companies profit. online - kaggle. https://www.kaggle.com/datasets/rupakroy/1000-companies-profit
2. Bouchard, G., Trouillon, T., Perez, J., Gaidon, A.: Online learning to sample. arXiv preprint arXiv:1506.09016 (2015)
3. Chicco, D., Warrens, M.J., Jurman, G.: The coefficient of determination R-squared is more informative than SMAPE, MAE, MAPE, MSE and RMSE in regression analysis evaluation. PeerJ Comput. Sci. **7**, e623 (2021)
4. Cormode, G., Shkapenyuk, V., Srivastava, D., Xu, B.: Forward decay: a practical time decay model for streaming systems. In: 2009 IEEE 25th International Conference on Data Engineering, pp. 138–149. IEEE (2009)
5. Cortez, P., Silva, A.: Using data mining to predict secondary school student performance. In: Brito, A., Teixeira, J. (eds.) Proceedings of 5th FUture BUsiness TEChnology Conference (FUBUTEC 2008), pp. 5–12. EUROSIS, Porto, Portugal (2008)
6. Crammer, K., Dekel, O., Keshet, J., Shalev-Shwartz, S., Singer, Y.: Online passive aggressive algorithms. Online passive aggressive algorithms (2006)

7. Drucker, H., Burges, C.J., Kaufman, L., Smola, A., Vapnik, V.: Support vector regression machines. In: Advances in Neural Information Processing Systems, vol. 9 (1996)
8. Fontenla-Romero, Ó., Guijarro-Berdiñas, B., Martinez-Rego, D., Pérez-Sánchez, B., Peteiro-Barral, D.: Online machine learning. In: Efficiency and Scalability Methods for Computational Intellect, pp. 27–54. IGI Global (2013)
9. Jothimurugesan, E., Tahmasbi, A., Gibbons, P., Tirthapura, S.: Variance-reduced stochastic gradient descent on streaming data. In: Advances in Neural Information Processing Systems, vol. 31 (2018)
10. Kasuya, E.: On the use of R and R squared in correlation and regression. Technical report, Wiley Online Library (2019)
11. Loshchilov, I., Hutter, F.: Decoupled weight decay regularization. arXiv preprint arXiv:1711.05101 (2017)
12. Maulud, D., Abdulazeez, A.M.: A review on linear regression comprehensive in machine learning. J. Appl. Sci. Technol. Trends $1(4)$, 140–147 (2020)
13. Mohri, M., Rostamizadeh, A., Talwalkar, A.: Foundations of Machine Learning. MIT Press, Cambridge (2018)
14. Ozer, D.J.: Correlation and the coefficient of determination. Psychol. Bull. $97(2)$, 307 (1985)
15. Prasad, S., Bruce, L.M.: Decision fusion with confidence-based weight assignment for hyperspectral target recognition. IEEE Trans. Geosci. Remote Sens. $46(5)$, 1448–1456 (2008)
16. Ruder, S.: An overview of gradient descent optimization algorithms. arXiv preprint arXiv:1609.04747 (2016)
17. Smola, A., Vishwanathan, S.: Introduction to Machine Learning. Cambridge University Press, Cambridge (2008)
18. Vapnik, V.: The Nature of Statistical Learning Theory. Springer, New York (1999). https://doi.org/10.1007/978-1-4757-3264-1
19. van Wieringen, W.N.: Lecture notes on ridge regression. arXiv preprint arXiv:1509.09169 (2015)
20. Abu-Mostafa, Y.S., Magdon-Ismail, M., Lin, H.T.: Learning from data (2012). https://amlbook.com/
21. Younger, A.S., Conwell, P.R., Cotter, N.E.: Fixed-weight on-line learning. IEEE Trans. Neural Netw. $10(2)$, 272–283 (1999)

Applied Data Science, Artificial Intelligence, and Data Engineering

Dimension Reduction Stacking for Deep Solar Wind Clustering

Daniel T. Carpenter[1] , Henry Han[2] , and Liang Zhao[1(✉)]

[1] Department of Climate and Space Sciences and Engineering,
University of Michigan, Ann Arbor, MI 48109, USA
{dcar,lzh}@umich.edu
[2] Department of Computer Science, School of Engineering and Computer Science,
Baylor University, Waco, TX 76798, USA
henry_han@baylor.edu

Abstract. In-situ observations of solar wind plasma exhibit statistical differences according to their coronal origins. These in-situ conditions are a direct result of various processes such as ionization and acceleration occur in the inner corona. Machine learning methods have been successful in characterizing solar wind in-situ observations using unsupervised deep clustering and dimensionality reduction techniques, but it remains unclear as to how solar wind data embedding and downstream clustering could be improved while providing better interpretability in machine learning process. In this study, we explore the impact of distance metrics on solar wind in-situ data clustering. We evaluate the metric performance by applying it to dimension-reduction-stacking and deep clustering techniques and comparing it with state-of-the-art methods using solar wind in-situ measurements. Our work demonstrates the potential for customized distance metrics to improve the interpretability and performance of deep clustering approaches applied in solar wind in-situ observations.

Keywords: Solar wind · Classification · Machine Learning · Heliophysics

1 Introduction

The solar wind is the stream of supersonic ionized particles released from the Sun, and drives *space weather* at Earth's geo-space environment. Space weather impacts Earth's climate, satellite communication, power grids, and other domains important to life on the surface. The physical processes occurring in the base of the solar corona that ionize, heat, and accelerate the solar wind plasma are of central importance to space weather forecasting and the ways in which the Sun affects the Earth.

One way we observe the solar wind is through in-situ measurements (e.g., ACE and Ulysses missions), which are inextricably tied to conditions in the solar corona–the outermost part of the sun's atmosphere–where the solar wind

H. Han and E. Baker (Eds.): SDSC 2023, CCIS 2113, pp. 111–125, 2024.
https://doi.org/10.1007/978-3-031-61816-1_8

originates. The in-situ properties of solar wind can be linked back to the coronal structures where it originates [24]. These properties encapsulate the dynamic and thermal properties of the bulk solar wind plasma (protons), minor constituents (Helium and heavier ions, such as Carbon, Nitrogen, Oxygen, Neon, Magnesium, Silicon, Sulfur, Iron, etc.), and magnetic field associated with them. Properties associated with these so-called heavy ions include total abundances and charge states, either as the average charge state of a specific element, or by the ratio of densities of individual charge states (such as O^{7+}/O^{6+}, C^{6+}/C^{5+}, measured by SWICS instrument onboard ACE and Ulysses).

There are certain solar wind structures identifiable by in-situ measurements. For example, plasma originating from coronal hole (CH) regions is usually observed to have high proton speeds: the aptly named fast wind. The plasma from CHs also has lower charge state ratios, indicating cooler electron temperatures and low plasma densities in the coronal origins [4,26]. This is in contrast of the slow solar wind, whose coronal origins are more difficult to ascertain from in-situ properties. The slow solar wind could come from anywhere from the periphery of active regions (ARs) [11,13], the helmet streamers [21,22], or the pseudostreamers [5,19,23], and so forth. The wind from these source regions is more ionized, which indicates hotter electron temperatures and higher densities in their coronal sources. Solar wind can also be attributed to occasional transient events which are important to space weather, such as Interplanetary Coronal Mass Ejections (ICMEs), which are energetic eruptions originating in magnetically active regions [7,8].

The Challenges of Solar Wind Classification. The solar wind has been traditionally classified according in-situ physical properties via statistical means; however, there are at least three challenges that arise when attempting to use in-situ properties to assign different types of solar wind to specific coronal sources: 1) singular observable, such as proton speed, is of poor use as a categorization metric (slow wind can arise from CHs [2,9,17,20,22]); 2) in-situ solar wind speeds and composition are on a continuum [27], and 3) the dimensionality of the data limits how the behavior can be visualized (there are 77 different parameters related to the heavy ion composition and elemental abundances alone ACE/SWICS [6]).

Unsupervised learning methods and dimensionality reduction algorithms have already proven effective at answering these challenges as data-driven characterization schemes [1,3,15]. However, an approach has yet to be shown which minimizes instances of subjectivity in parameter selection and explains how the downstream embedding and clustering results are delivered. In this work, we detail the appropriate domain knowledge for solar wind data and introduce a novel deep clustering approach, PCA+t-SNE+DBSCAN, for characterizing the solar wind using in-situ properties. This method utilizes our dimension reduction stacking technique, PCA+t-SNE, proposed in this study, along with various distance metric probing, to effectively and more transparently identify solar wind clusters. To the best of our knowledge, this is the first explainable machine learning method proposed for the clustering of solar wind data. The proposed dimension reduction stacking can also be extended to other AI and data science fields.

2   Dimension Reduction Stacking

The application of dimension reduction stacking is novel to the field of solar physics. Dimension reduction stacking is the technique of combining reduction methods by using the output of one method as the input of another. Formally, the original input data $X = \{x_i\}_{i=1}^{N}$, where $x_i \in \mathbb{R}^M$, is mapped to a low-dimensional representation $Y = \{y_i\}_{i=1}^{N}$, where $y_i \in \mathbb{R}^l$ and $l \ll p$. The stacking results in a composite function $f(g(X)) \rightarrow Y$. Typically, f and g belong to different types of dimension reduction methods, and their combination enables the extraction of features at a deeper level by having one method address the limitations that another has in the stack.

2.1   Principal Component Analysis

The first dimension reduction method we used in our stack is Principal Component Analysis (PCA). PCA focuses on minimizing information loss by creating new uncorrelated variables that successively maximize variance. These variables are called principal components (PCs), and they are the solutions to an eigenvalue/eigenvector problem of the covariance matrix of the input data. The quality of the reduction can be measured using the variability associated with the set of retained PCs. To measure the quality the amount of selected PC's, the cumulative explained variance percentage is calculated. PCA, being a classic holistic method, is capable of extracting the global behavior of the data; however, PCA usually cannot capture the local behavior of the data because each PC only contains some levels of global characteristics of the data.

2.2   t-Distributed Stochastic Neighboring with Embedding

The second dimension reduction technique in the stack is the t-Distributed Stochastic Neighboring with Embedding (t-SNE). This is a non-linear method, capable of capturing local data behaviors in the dimensionality reduction. The t-SNE algorithm minimizes the Kullback-Leiber (K-L) divergence between a distribution P and student t-distribution Q to achieve the low-dimensional embedding by solving a non-convex optimization problem. The Gaussian and t-distributions model the pairwise similarities between data points in the original high-dimensional input space and embedding. t-SNE aims to force a similarity of the embedding data to the original data by seeking the minimum K-L distance between P and Q. By having the PCA transformation emphasize the global structure of the data, and t-SNE generate an embedding that captures local structure in the data (in multiple scales), the data expression can be represented in two dimensions in a way that considers both global and local structures. Once the data is in this form, the embedding can be clustered in order to identify attributes of the clusters.

2.3   PCA+t-SNE Dimension Reduction Stacking

The dimension reduction stacking is done by composing PCA and t SNE, titled PCA+t-SNE. We project the input data onto the PCA space while ensuring a minimum of 95% total explained variance. Subsequently, reduction is used to compute distance matrices. These distance matrices will then be fed into t-SNE. It is worth mentioning that PCA+t-SNE outperforms PCA+UMAP in terms of performance, as t-SNE has a better ability to capture local behavior of the data compared to UMAP [10]. This observation is further supported by our downstream DBSCAN clustering results.

Algorithm 1: Dimension Reduction Stacking (X,η,p)

---

1 **Input:**
2   The input data: $X \in \mathbb{R}^{N \times M}$.
3   The explained variance ratio: η.
4   The perplexity in t-SNE: p.
5 **Output:**
6   The dimension reduction stacking embedding: $X_{PCA+TSNE}$.
7 **Begin.**   PCA for the scaled data:
8      $X_{PCA}, pcVariance \leftarrow pca(X)$.
9   Retrieve reduction:
10 **IF** $\sum_{i=1}^{i=l} pcVariance_i \geq \eta$
11      $X_{embedding} \leftarrow X_{PCA}[:,1:l]$.
12 Compute pairwise distances:
13      $D_X \leftarrow f_{dist}(X_{embedding})$
14 t-SNE embedding:
15      $X_{PCA+TSNE} \leftarrow tsne(D_X,p)$
16 **Return:** $X_{PCA+TSNE}$.
17 **End**

---

3   Data and Preprocessing

We use the Advanced Composition Explorer (ACE) spacecraft as the platform for our data. ACE is positioned at the L1 point, measuring the solar wind plasma and interplanetary magnetic field since 1998. A subset of the data from 2000–2002 was chosen, as because the heightened solar activity in this time range resulted in more frequent equatorial CHs and ICMEs, providing in a more balanced inclusion of solar wind sources in this interval. From this range, we use a random subset of 2500 2-hour-binned measurements.

To create the data expression for reduction, we collect 12 variables linked to solar wind in-situ signatures: from the Solar Wind Electron, Proton, and Alpha

Monitor (SWEPAM) [14], we use the proton temperature (T_p) and proton density (n_p) to compute the proton entropy $(S_p = T_p \, n_p^{-2/3})$, and we also include alpha-to-proton ratio (denoted as α/H). From SWICS [9], we include the elemental composition (relative abundances of Magnesium, Silicon, Iron, Carbon, Neon, and Helium to Oygen, denoted as Mg/O, Si/O, Fe/O, C/O, Ne/O, and He/O respectively) and heavy ion composition signatures of Oxygen, Carbon and Iron $(O^{7+}/O^{6+}, \, C^{6+}/C^{4+}, \, C^{6+}/C^{5+}, \, \text{and} \, \langle Q_{Fe} \rangle)$.

After removing samples with null values in the original data set and taking the random subset (as described previously), the data variables (described above) are scaled using a Min-Max scaling along each dimension.

4 Solar Wind Deep Clustering Under Dimension Reduction Stacking

We generate a meaningful embedding space for input solar wind data that reveals both latent global and local data characteristics through PCA+tSNE dimension stacking, before seeking meaningful similarity via density-based clustering. PCA is applied as the first dimensionality reduction technique. The left panel of Fig. 1 shows the percentage of explained variance (blue) and accumulated explained variance (red) by each PC. The subset of PCs up to PC_7 explains 95% of the original data variance. All of the data are visualized in the first two PC components frame through a Gaussian Kernel Density Estimate (KDE) in the right panel of Fig. 1.

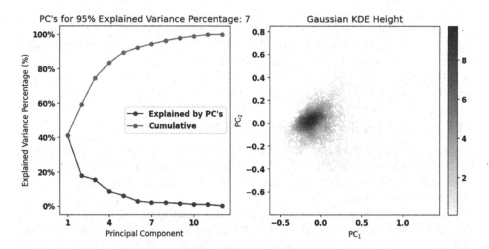

Fig. 1. Result of PCA on the in-situ solar wind data. The left panel shows the percentage of variance (blue) and cumulative sum (red) explained by each PC. The middle panel shows the eigenvalues of each PC. The right plot visualizes the Gaussian KDE height of the first two PCs. (Color figure online)

In our implementation, we utilize the projected data of the original solar wind dataset in the PCA subspace, maintaining a 95% explained variance ratio, for calculating pairwise distance matrices using the Euclidean, Cosine, and Mahalanobis distance metrics respectively. These pre-computed distance matrices are then employed as inputs for t-SNE to generate the t-SNE embedding. This PCA+tSNE dimension reduction stacking approach captures both the global and local characteristics of the data in dimension reduction, in addition to a denoising procedure. Moreover, this stacking method yields a meaningful embedding space for exploring similarities in solar wind data, which benefits subsequent density-based clustering, such as DBSCAN. DBSCAN is robust to noise and adaptable to data of any shape, making it suitable for the noisy, nonlinear solar wind data that can demonstrate any shape after dimension reduction stacking. Deep: PCA+t-SNE+DBSCAN here means an in-depth exploration of latent global and local data characteristics revealed in the latent embedding generated from PCA+t-SNE stacking, as well as the examination of various similarity metrics in PCA+t-SNE stacking.

DBSCAN (Density Based Spatial Clustering of Applications with Noise) is a density-based clustering algorithm designed to cluster data of arbitrary shape, and to account for noise. DBSCAN classifies points as either core, reachable, or outlier (noise). Core points are within a radius ε of a neighborhood–the size of the neighborhood is specified by a minimum number of points, including the point in question. A reachable point is within radius ε of a core point, but does not neighbor the required minimum number of points. An outlier is a point which is beyond ε of any core point. In our context, outliers will be solar wind samples which have physical qualities which differ enough from the main groups of solar wind. The core points form clusters because of their high densities, while the reachable points form the edge of said clusters and the outliers stand out as noise. DBSCAN will allow us to find the boundaries of clusters without imposing any model or numerical restrictions. Once these cluster labels are generated, we can then project the labels back onto the original data in order to examine the underlying physics of the solar wind clusters.

4.1   Deep Clustering Under Different Distance Metrics

We employ three distance metrics-Euclidean, cosine, and Mahalanobis-in PCA+t-SNE stacking before DBSCAN clustering. This implies that these distances are utilized to calculate the pairwise distance matrix using data projected into the PCA subspace

The corresponding PCA+t-SNE+DBSCAN clustering results are shown in Figs. 2, 3, and 4. In the DBSCAN clustering process, we set the minimum number of points for a cluster to be considered a core cluster to 75 points. We then vary epsilon for each embedding until we see some of the smaller scale clusters. The epsilon values are shown in Table 1. The clusters are projected onto physical parameters associated with each data point.

The first distance metric we examine is the Euclidean metric, resulting in 6 clusters and one outlier group (Fig. 2). The winds in Class 2 have very low

Table 1. Selected epsilon values for each distance metric

Distance Metric	Euclidean	Cosine	Mahalnobis
Epsilon	11.0	10.3	11.2

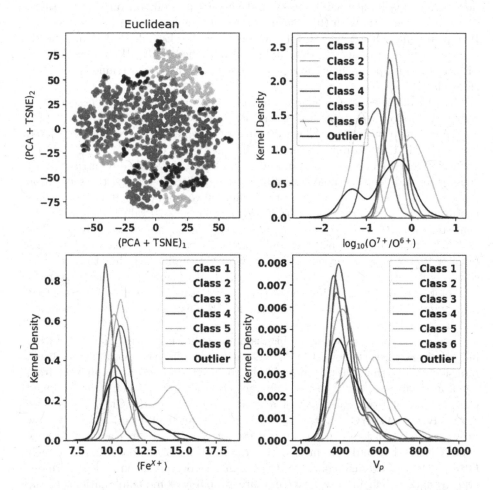

Fig. 2. Embedding and clustering using Euclidean distance as the metric in the PCA+TSNE process (top left). Class labels are assigned by DBSCAN, and are projected onto the charge state ratio of oxygen (O^{7+}/O^{6+} ratio), the average charge state of Iron, and the bulk proton speed (top right, bottom left, and bottom right, respectively).

O^{7+}/O^{6+} and relatively high V_p, indicative of equatorial coronal hole associated fast wind. Class 3, 4, and 6 have relatively high O^{7+}/O^{6+} and slow proton speed, indicative of typical streamer-associated slow wind [22]. The O^{7+}/O^{6+} of 0.145 (logarithm base 10 value is about -0.84) was found by previous work [22] that can effectively separate the streamer-associated slow solar wind from the coronal-hole-associated fast wind; and here we see that the coronal hole (cluster 2) and slow wind (clusters 3, 4, 6) are separated by a very similar value of O^{7+}/O^{6+}. Class 5 is an interesting class; it has characteristics similar to slow solar wind; however, it has higher wind speeds and very high average iron charge states. These are consistent with the characteristics of ICMEs passing by a spacecraft. It appears that clustering the solar wind using this Euclidean distance metric can be useful to extract ICMEs, and separated the non-ICME solar wind into distinct sub-groups, such as coronal-hole associated fast and streamer-associated slow wind.

The result based on Cosine distance metric is shown in Fig. 3. There are two very similar slow wind clusters (2 and 6), possessing relatively high O^{7+}/O^{6+} ratio and slow proton speed. These two clusters are likely the streamer-associated slow solar wind. Oppositely, in cluster 5 the solar winds have low O^{7+}/O^{6+} ratios and relatively high proton speed, those winds are more likely contributed by the low latitude coronal-hole associated fast wind. Interestingly, the separation point of O^{7+}/O^{6+} between these two different coronal-originated solar winds is still located near the value of 0.145 (logarithm base 10 value is about -0.84), consistent with the result of Euclidean distance and the previous study [22]. The O^{7+}/O^{6+} ratios of the clusters 1 and 3 are in between the streamer slow wind (cluster 2 and 6) and coronal-hole fast wind (5), indicating that they may be some combination of these two types of winds. In the average charge state of Iron plot, the relatively high value of Iron charge state in the class 4 implies that some winds in this class may contain the ICME plasma, but not all of them are ICMEs.

The result based on Mahalanobis distance is shown in Fig. 4. The solar wind in class 6 possesses the lowest O^{7+}/O^{6+} ratio, and relatively high proton speed, indicating that this class is more likely to be the fast wind originated from equatorial coronal holes. Class 5 is characterized as having the highest averaged charge state of Iron, indicative of a group of ICME winds. Class 3 possesses relatively high O^{7+}/O^{6+} ratio (mostly higher than 0.145) and relatively slow proton speed, consistent with the features of typical streamer-associated slow wind. Classes 1, 2 and 4 seem to posses the majority of the data, however their O^{7+}/O^{6+} ratio, average charge state of Iron and proton speed are in the moderate ranges which implies that they are probably some combinations of the coronal-hole and streamer winds, therefore they cannot be assigned to any specific solar wind types or coronal origins.

Maximum Fusion Distance Metric. The clustering results calculated using the three different distance metrics have different strength and weakness in relating them with the solar wind features and the coronal origins. It is hard to determine which distance metrics is the best one among these three. Therefore, in the next step, we create a new distance metric which is designed to prioritize the effectiveness of all of the three metrics, in order to obtain the optimized solar wind clustering result.

The new distance metric is defined as the maximum of the three normalized metrics as calculated previously (Eq. 1). We name this new distance metric Maximum Fusion.

$$p_{max} = \max(\frac{p_{euc}}{\max(p_{euc})}, \frac{p_{cosine}}{\max(p_{cosine})}, \frac{p_m}{\max(p_m)}) \qquad (1)$$

We then apply the same t-SNE parameters and clustering process, retaining MinPts as 75 and choosing epsilon to reveal multiple scales of clusters. The result of the clustering is shown in Fig. 5. It is clear that the solar wind clusters are in general separated at the threshold of $O^{7+}/O^{6+} = 0.145$ (logarithm base 10 value is -0.84), with class 2 and 5 as hotter (O^{7+}/O^{6+} ratio > 0.145, streamer-

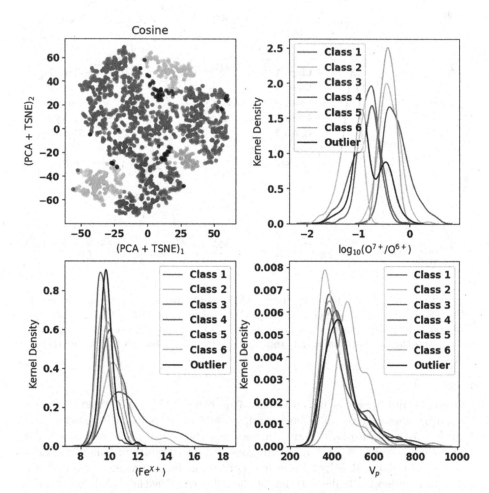

Fig. 3. Embedding and clustering using Cosine distance as the metric in the PCA + TSNE process (top left). Class labels are assigned by DBSCAN, and are projected onto the charge state ratio of oxygen (O^{7+}/O^{6+} ratio), the average charge state of Iron, and the bulk proton speed (top right, bottom left, and bottom right, respectively).

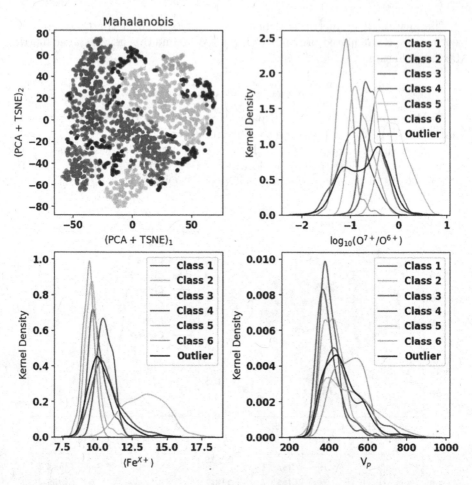

Fig. 4. Embedding and clustering using Mahalanobis distance as the metric in the PCA+TSNE process (top left). Class labels are assigned by DBSCAN, and are projected onto the charge state ratio of oxygen (O^{7+}/O^{6+} ratio), the average charge state of Iron, and the bulk proton speed (top right, bottom left, and bottom right, respectively).

associated) and slower wind and class 3 as colder (O^{7+}/O^{6+} ratio < 0.145, coronal-hole associated) and faster wind; meanwhile class 1 which possesses the majority of the data points, stays in between. Particularly, the winds in class 4 have the highest O^{7+}/O^{6+} ratios and average charge state of Iron, indicating that they are likely ICMEs. These five clusters in Fig. 5 are more explainable and better understandable in terms of the solar wind features and the coronal origins than the results from the other three distance metrics.

5   Conclusion

In this work, we find that PCA+t-SNE can characterize in-situ solar wind data in a way that reveals the hidden features in the data. We then present the solar wind clustering results using DBSCAN with three distance metrics, Euclidean, Cosine, and Mahalanobis, and a new distance metrics that we invent, Max Fusion. We describe the impact that these metrics have on the solar wind clustering, and compare the clustering results (Table 2).

Table 2. Comparison of the solar wind clustering results among different distance metrics

	slow wind (high O^{7+}/O^{6+})	fast wind (low O^{7+}/O^{6+})	CME (high O^{7+}/O^{6+} and Q_{Fe})	undefined
Euclidean	3 classes (3,4,6)	1 class (2) wide speed range	1 class (5)	1 class (1)
Cosine	2 classes (2,6)	1 class (5) wide speed range	not very clear	2 classes (1,3)
Mahalanobis	1 class (3)	1 class (6) wide speed range	1 class (5)	3 classes (1,2,4)
Max Fusion	2 classes (2,5)	1 class (3)	1 class (4)	1 class (1)

The comparison of the clustering results show that all of the four distance metrics can produce solar wind clusters that are indicative of streamer-associated slow solar wind (slow speed and high O^{7+}/O^{6+} ratio). Euclidean and Mahalanobis distance metrics can also produce a cluster of solar wind that matches the characteristics of ICMEs (high O^{7+}/O^{6+} and average charge state of Iron), but Cosine distance metrics fails to produce a cluster of solar wind that can be exclusively considered as ICMEs. All of the first three distance metrics result in a cluster of solar wind that possesses low O^{7+}/O^{6+} ratio solar wind, and has a proton speed range spanning from 300 to 700 Km/s with two peaks, indicating that this cluster may be partially contributed by coronal-hole associated winds, but also with some contamination from other slow-speed types of winds. Differently, the clusters identified using Max Fusion distance metrics show clear different types of solar wind: two clusters are slow, streamer-associate wind; one cluster is coronal-hole associate fast wind; and one cluster is like ICME wind. Therefore, we conclude that the Max Fusion distance metrics so far is the best distance metrics that can effectively classify the solar wind into different categories of physically distinct characteristics and indicative of different coronal origins. We summarize the comparison of these clustering results in Table 2.

6   Discussion

The MinPts parameter is essential in defining core points and influencing cluster density in the DBSCAN algorithm, particularly for large, noisy datasets such as

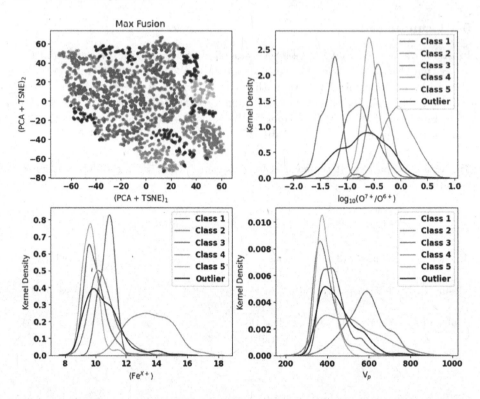

Fig. 5. Embedding and clustering using max-fusion distance as the metric in the PCA+TSNE process (top left). Class labels are assigned by DBSCAN, and are projected onto the charge state ratio of oxygen (O^{7+}/O^{6+} ratio), the average charge state of iron, and the bulk proton speed (top right, bottom left, and bottom right, respectively).

solar wind data. This becomes even more significant when applied to embedding data derived from the proposed dimension reduction stacking (PCA+tSNE). Our study, as shown in the left panel of Fig. 6, illustrates how the Eps (epsilon) value fluctuates in relation to the Euclidean distance metric in the embedding. Teach-nically, it refers to the radius of a neighborhood around a solar wind projection point in the embedding space.

A lower MinPts setting results in a restricted epsilon range and a higher cluster count, whereas a very high MinPts leads to underfitting, as indicated by a reduced number of clusters. Interestingly, the left side of these curves tends to display a consistent cluster count across various MinPts values, though this can cause overfitting with many outliers. To counteract this, a higher MinPts value enables DBSCAN to concentrate on fewer, larger-scale clusters, a crucial approach in our study to prevent the formation of overly small clusters and ensure the applicability of our findings to a wide range of solar wind datasets. Therefore, we have selected a MinPts value of 75 for our analysis. However, it's important to note that this empirical choice may not be the best fit for generalization in larger solar wind datasets.

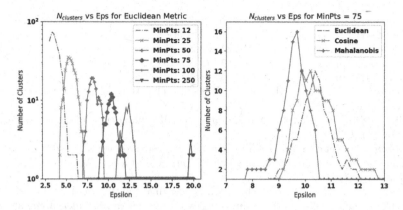

Fig. 6. The number of clusters that result from DBSCAN clustering with increasing epsilon. The left panel shows varying MinPts on the Euclidean distance metric. The right panel shows various distance metrics on MinPts = 75.

Moreover, our previous findings indicate that the epsilon values are relatively similar across different distance metrics. As depicted in the right panel of Fig. 6, there is a noticeable overlap in cluster numbers when using both Euclidean and Cosine distance metrics, in contrast to the Mahalanobis metric, which shows peak values at lower epsilon levels. This observation suggests that the choice of epsilon not only depends on the scale of the structures under study but also varies in its impact across different distance metrics within the embedding space. To address this, we are developing an optimal parameter tuning strategy for large-scale solar wind data analysis. This approach involves utilizing various distance metrics and a probing learning method that leverages a small subset of the data for initial insights. Additionally, we are incorporating distance metric learning techniques, including Siamese and Triplet Networks, to uncover more meaningful and insightful solutions in our analysis [28].

Acknowledgements. The work of D.C. is supported by NASA grant 80NSSC 22K1015. L.Z. is supported by NASA Grants 80NSSC21K0579, 80NSSC22K1015, NSF SHINE grant 2229138, and NSF Early Career grant 2237435. H.H is supported by the McCollum endowed chair startup fund of Baylor University.

References

1. Bloch, T., Watt, C., Owens, M. et al.: Data-driven classification of coronal hole and streamer belt solar wind. Sol. Phys. **295**(41) (2020) https://doi.org/10.1007/s11207-020-01609-z
2. Bravo, S., Stewart, G.A.: Fast and slow wind from solar coronal holes. Astrophys. J. **489**, 992 (1997)
3. Carpenter, D., Zhao, L., Lepri, S.T., Han, H.: Characterizing in-situ solar wind observations using clustering methods. In: Han, H., Baker, E. (eds.) SDSC 2022. CCIS, vol. 1725, pp. 125–138. Springer, Cham (2022). https://doi.org/10.1007/978-3-031-23387-6_9

4. Cranmer, S.R.: Coronal holes and the high-speed solar wind. Space Sci. Rev. **101**, 229 (2002)
5. Crooker, N.U., Antiochos, S.K., Zhao, X., Neugebauer, M.: Global network of slow solar wind. J. Geophys. Res. **117**, A04104 (2012). https://doi.org/10.1029/2011JA017236
6. Garrard, T., Davis, A., Hammond, J., et al.: The ACE science center. Space Sci. Rev. **86**, 649–663 (1998)
7. Gibson, S.E., Fan, Y.: The partial expulsion of a magnetic flux rope. Astrophys. J. **637**(1), L65–L68 (2006)
8. Gibson, S.E., Fan, Y., Török, T., Kliem, B.: The evolving sigmoid: evidence for magnetic flux ropes in the corona before, during, and after CMEs. Space Sci. Rev. **124**(1–4), 131–144 (2006)
9. Gloeckler, G., Cain, J., Ipavich, F.M., et al.: Investigation of the composition of solar and interstellar matter using solar wind and pickup ion measurements with SWICS and SWIMS on the ACE spacecraft. Space Sci. Rev. **86**, 497 (1998)
10. Han, H., Wentian, L., Wang, J., Qin, G., Qin, X.: Enhance explainability of manifold learning. Neurocomputing **500**, 877–895 (2022)
11. Ko, Y.-K., Raymond, J.C., Zurbuchen, T.H., et al.: Abundance variation at the vicinity of an active region and the coronal origin of the slow solar wind. Astrophys. J. **646**, 1275 (2006)
12. Lepri, S.T., Zurbuchen, T.H.: Iron charge state distributions as an indicator of hot ICMEs: possible sources and temporal and spatial variations during solar maximum. J. Geophys. Res. **109**, A01112 (2004). https://doi.org/10.1029/2003JA009954
13. Liu, S., Su, J.T.: Multi-channel observations of plasma outflows and the associated small-scale magnetic field cancellations on the edges of an active region. Astrophys. Space Sci. **351**, 417 (2014)
14. McComas, D., Bame, S., Barker, P., et al.: Solar wind electron proton alpha monitor (SWEPAM) for the advanced composition explorer. Space Sci. Rev. **86**, 563–612 (1998)
15. Roberts, D.A., et al.: Objectively determining states of the solar wind using machine learning. ApJ **889**, 153 (2020)
16. Smith, C., L'Heureux, J., Ness, N., et al.: The ACE magnetic fields experiment. Space Sci. Rev. **86**, 613–632 (1998)
17. Stakhiv, M., Landi, E., Lepri, S.T., Oran, R., Zurbuchen, T.H.: On the origin of mid-latitude fast wind: challenging the two-state solar wind paradigm. Astrophys. J. **801**, 100 (2015)
18. Wang, Y.-M., Ko, Y.-K.: Observations of slow solar wind from equatorial coronal holes. Apj **880**, 146 (2019)
19. Wang, Y.-M., Grappin, R., Robbrecht, E., et al.: On the nature of the solar wind from coronal pseudostreamers. Astrophys. J. **749**, 182 (2012)
20. Wang, Y.-M., Ko, Y.-K., Grappin, R.: Slow solar wind from open regions with strong low-coronal heating. Astrophys. J. **691**, 760 (2009)
21. Zhao, L., Landi, E., Zurbuchen, T.H., Fisk, L.A., Lepri, S.T.: The evolution of 1 AU equatorial solar wind and its association with the morphology of the heliospheric current sheet from solar cycles 23 to 24. Astrophys. J. **793**, 44, 8 pp (2014). https://doi.org/10.1088/0004-637X/793/1/44
22. Zhao, L., Zurbuchen, T.H., Fisk, L.A.: Global distribution of the solar wind during solar cycle 23: ACE observations. Geophys. Res. Lett. **36**, L14104 (2009)
23. Zhao, L., Gibson, S.E., Fisk, L.A.: Association of solar wind proton flux extremes with pseudostreamers. J. Geophys. Res. Space Phys. **118**, 2834–2841 (2013)

24. Zhao, L., et al.: On the relation between the in-situ properties and the coronal sources of the solar wind. Astrophys. J. **846**(2), 135 (2017)

25. Zhao, L., et al.: An anomalous composition in slow solar wind as a signature of magnetic reconnection in its source region. ApJS **228**, 1 (2017)

26. Zirker, J.B.: Coronal holes and high-speed wind streams. Rev. Geophys. Space Phys. **15**, 257 (1977)

27. Zurbuchen, T.H., Fisk, L.A., Gloeckler, G., von Steiger, R.: The solar wind composition throughout the solar cycle: a continuum of dynamic states. Geophys. Res. Lett. **29**(9) (2002). https://doi.org/10.1029/2001GL013946

28. Han, H., Li, D., Liu, W., Zhang, H., Wang, J.: High dimensional mislabeled learning. Neurocomputing **573**, 127218 (2024)

An Experimental Study of the Joint Effects of Class Imbalance and Class Overlap

Yutao Fan[1,2,3,4(✉)], Heming Huang[1,2,3], CaiRang DangZhi[1,2,3], XiaWu Ji[1,2,3], and Qian Wu[1,2,3]

[1] Qinghai Normal University, Xining 810008, China
fanyutao2005@126.com, huanghm@qhnu.edu.cn
[2] State Key Laboratory of Tibetan Intelligent Information Processing and Application, Xining 810008, China
[3] Key Laboratory of Tibetan Information Processing, Ministry of Education, Xining 810008, China
[4] North China Institute of Science and Technology, Beijing 065201, China

Abstract. It has been pointed out that the class imbalance problem is one of the critical areas in classification. Furthermore, existing literatures show that other factors such as class overlap, small disjuncts, and noises will aggravate classification performance when they are combined with class imbalance. In this work, we focus on the joint effects of class imbalance and class overlap, and study binary classification performances of six algorithms under different combination of imbalance ratios and overlap degrees. The experiments corroborate that different types of classifiers show distinct robustness to class imbalance and overlap degree. We arrive the conclusion that essentially the densities of different regions of data space affect the classification performance. In addition, based on observations from our experiments, we infer to changing the densities of different regions of data space should be a good way to address problem of class imbalance and class overlap.

Keywords: Class Imbalance · Class Overlap · Classification · Probability Density

1 Introduction

Class imbalance remains one of the most challenging problems in machine learning [1–3]. For a binary classification scenario, class imbalance means one class called as majority class has a large number of samples while the other class called as minority class has only much fewer samples compared with the majority class. Because of skewed distributions of two classes, traditional algorithms usually bias towards the majority class and even totally omit the minority class, thus deteriorating classification performance, especially for the minority class [4].

In fact, many researches prove that class imbalance does not bring any impacts on classification performance if the two classes are linearly separable [5]. Therefore, it is

H. Han and E. Baker (Eds.): SDSC 2023, CCIS 2113, pp. 126–140, 2024.
https://doi.org/10.1007/978-3-031-61816-1_9

obvious that class imbalance only poses bad impacts on classification performance in the cases of non-linearly separable distributions of two classes.

Class overlap is often defined that there are some regions of data space where a similar number of samples from each class is contained [6–8]. Especially, in the case of binary classification, class overlap occurs when some of the minority samples are intertwined with some majority ones in some regions of data space, making it difficult or impossible to distinguish between classes accurately. Class overlap makes linearly separable situation impossible and if combined with class imbalance the classification performance will be worse.

In this work, we focus on the study of the behaviors of six classifiers under the presence of both class imbalance and class overlap in the scenario of binary classification. In order to fully control all the variables that we want to analyze, we create two artificial datasets, which conform with Gaussian distribution and uniform distribution, respectively. We change those data difficulty factors like class distribution density and sample size of two classes, etc. in order to unveil how these factors will affect the classification performance when class imbalance and class overlap do not occur at the same time or even occur simultaneously. Specifically, this study answers the following research questions:

1. Under the non-linearly separable circumstance, what impacts do changes of standard deviation abbreviated as STD and IR bring to classification performance?
2. Under the non-linearly separable case, to what extent do different parts of minority class like high density parts, sparse regions, small disjuncts, and noise samples degrade the classification performance?
3. Which classifiers perform best or worst in different cases where class imbalance and class overlap occur at the same time or not?

We conduct rich experiments with artificial datasets and use some metrics including specificity, recall/sensitivity, precision, F1 score, and balanced accuracy to assess the classification performance.

2 Machine Learning Algorithms

In this section, we describe the classifiers or machine learning algorithms used in the following experiments briefly. The six classifiers used belong to two families whose primary differences are their learning strategies, that is, local learning or global learning [9]. The aim is to find performance differences among different classifiers under different data sets with combination of class imbalance and class overlap. These six classifiers include a Nearest Neighbor classifier, a Naïve Bayes classifier, a MLP neural network, a SVM classifier, a LDA classifier and a random forest classifier. The first one belongs to the local learning algorithm and the other five are global ones.

2.1 Naïve Bayes Classifier

The Naïve Bayes classifier [10] is arguably one of the simplest probabilistic schemes, following from Bayes' theorem with the "naive" assumption of conditional independence between every pair of features given the value of the class variable. Bayes' theorem states the following relationship shown as Eq. (1), given class variable y and dependent feature vector x_1 through x_n.

$$P(y|x_1, \ldots, x_n) = \frac{P(y)P(x_1, \ldots, x_n|y)}{P(x_1, \ldots, x_n)} \tag{1}$$

Equation (1) can be simplified to Eq. (2) when the naïve conditional independence assumption, that is, $P(x_i|y, x_1, \ldots, x_{i-1}, x_{i+1}, \ldots, x_n) = P(x_i|y)$ is used.

$$P(y|x_1, \ldots, x_n) = \frac{P(y) \prod_{i=1}^{n} P(x_i|y)}{P(x_1, \ldots, x_n)} \tag{2}$$

Since $P(x_1, \ldots, x_n)$ is constant given the input, Eq. (3) can be obtained.

$$P(y|x_1, \ldots, x_n) \propto P(y) \prod_{i=1}^{n} P(x_i|y) \tag{3}$$

And then Maximum A posteriori estimation is used to estimate $P(y)$ and $P(x_i|y)$. Finally, $\hat{y}$ obtained as shown in Eq. (4) is the predicted label in classification.

$$\hat{y} = \underset{y}{\mathrm{argmax}} P(y) \prod_{i=1}^{n} P(x_i|y) \tag{4}$$

The different naive Bayes classifiers differ mainly by the assumptions they make regarding the distribution of $P(x_i|y)$[11]. There are many different types of naïve Bayes classifiers including but not limited to Gaussian Naïve Bayes, Multinomial Naïve Bayes and Bernoulli Naïve Bayes. In our experiments, Gaussian Naive Bayes is used.

2.2 Nearest Neighbor Classifier

The Nearest Neighbor Classifier [12] is a kind of supervised machine learning algorithm that operates based on spatial distance measurements. It has the characteristic of locality which means it can address data regions with different local data densities compared to most global classifiers like MLP and Naïve Bayes classifier. Usually, smaller values of k will guarantee its local nature. So, in our experiments, the value of k is 0.

2.3 Support Vector Machine

The Support Vector Machine, abbreviated as SVM, has been successfully applied to many classification problems in numerous domains and shows better generalization performance than other learning algorithms [13]. It constructs a hyper-plane or set of hyper-planes in a high or infinite dimensional space. SVCs short for Support Vector Classifiers are a popular choice for many machine learning applications due to their ability to achieve high accuracy while still being computationally efficient. The basic idea behind SVC is as follows. Given training vectors $x_i \in \mathbb{R}^p, i = 1, 2, \ldots n$, in two classes and a vector $y \in \{1, -1\}^n$, the problem classification is to find $\omega \in \mathbb{R}^p$ and $b \in \mathbb{R}$

so that the prediction given by $sign(\omega^T \phi(x)\_b)$ is correct for most samples. SVC tackles a convex optimization problem expressed as Eq. (5) to obtain ω and b.

$$\min_{\omega, b, \varsigma} \frac{1}{2} \omega^T \omega + c \sum_{i=1}^{n} \varsigma_i \tag{5}$$

$$subject\ to\ y_i \left(w^T \phi(x_i) + b \right) \geq 1 - \varsigma_i \varsigma_i \geq 0, i = 1, 2, \ldots n$$

To solve the optimization problem, SVC changes it into a dual problem expressed as Eq. (6).

$$\min_{\alpha} \frac{1}{2} \alpha^T Q \alpha - e^T \alpha \tag{6}$$

$$subject\ to\ y^T \alpha = 0 0 \leq \alpha_i \leq C, i = 1, 2, \ldots n$$

In Eq. (6), e is a vector of all ones and Q is an n by n positive semidefinite matrix, $Q_{ij} = y_i y_j K(x_i, x_j)$. $K(x_i, x_j) = \phi(x_i)^T \phi(x_i)$ is the kernel. α_i are the dual coefficients and are upper-bounded by C. Once the optimization problem is solved, the decision function for a given sample x becomes $\sum_{i \in SV} y_i \alpha_i K(x_i, x) + b$ and the predicted class correspond to its sign.

Different kernel functions can be specified for the decision function of SVMs so that linear and non-linear decision boundaries can be obtained. To some extent, non-linear decision boundaries benefit the classification performance more. In our experiments, the SVM with RBF kernel is used.

2.4 Linear Discriminant Analysis

Linear Discriminant Analysis (LDA) [14] is a technique used to find a linear combination of features that best separates the classes in a dataset. It is derived from simple probabilistic models which model the class conditional distribution of the data $P(X|y = k)$ for each class k. Predictions then are obtained by using Bayes' rule shown as Eq. (7), for each training sample $x \in R^d$.

$$P(y = k|x) = \frac{P(x|y = k)P(y = k)}{P(x)} = \frac{P(x|y = k)P(y = k)}{\sum_l P(x|y = l)P(y = 1)} \tag{7}$$

The log of $P(y = k|x)$ is obtained according to Eq. (7), as is shown in Eq. (8), where C_{st} corresponds to the denominator $P(x)$.

$$logP(y = k|x) = logP(x|y = k) + logP(y = k) + C_{st} \tag{8}$$

As for LDA, it is a special case of Eq. (8), that is, $P(x|y = k) = \frac{1}{(2\pi)^{d/2}|\sum_k|^{1/2}} \exp(-\frac{1}{2}(x - u_k)^t \sum_k^{-1} (x - u_k))$ and is expressed as Eq. (9).

$$logP(y = k|x) = -\frac{1}{2}(x - u_k)^t \sum_k^{-1} (x - u_k) + logP(y = k) + C_{st} \tag{9}$$

2.5 Multilayer Perceptron

The multilayer perceptron (MLP) neural network is an artificial neural network and hence, consists of interconnected neurons which process data through three or more layers. The basic structure of an MLP consists of an input layer, one or more hidden layers and an output layer, an activation function and a set of weights and biases. MLP is capable of handling both linearly separable and non-linearly separable data. There are several training algorithms for MLP. The most common is the backpropagation algorithm, which takes a set of training instances for the learning process [15]. In our experiments, two hidden layers are included and the relu function and adam optimizer are used respectively.

2.6 Random Forest

Random forest (RF) is a popular ensemble algorithm that provides fast training, interpretability, and excellent performance. Compared with other ensemble methods, random forest has some key benefits like reduced risk of overfitting desirable performance and feasibility, thus making it widely applied in many domains [16, 17].

3 Evaluation Metrics

The metric that is most widely used to evaluate a classifier performance in binary classification is overall accuracy. But when faced with class imbalance, the overall accuracy can only reflect the classification effect of the classifier as a whole, and cannot reflect the classification accuracy of the minority class, which is more important [18–21]. To evaluate the performance of the classifier more meaningfully, the form of the confusion matrix is often adopted.

Based on the confusion matrix, recall/sensitivity value and precision value which are calculated by Eq. (10) and Eq. (11) can be obtained. The precision value measures the classifier trustiness in classifying positive samples while the recall/sensitivity value measures how many positive samples are correctly classified by the classifier. In the case of imbalanced datasets, although recall/sensitivity is important, the accurate recognition of negative samples is also indispensable. Specificity calculated by Eq. (12) is used to calculate how many negative predictions made are correct. The use of recall/sensitivity and specificity means improvement on recall/sensitivity cannot be obtained through decreasing recognition of negative samples.

Anyhow, single metric is not comprehensive. Therefore, F1 scores, which are calculated by Eq. (13), are usually used to be interpreted as a harmonic mean of the precision and recall, especially in the case of imbalanced class distribution. Especially, the balanced accuracy score calculated by Eq. (14) is used to assess the classification performance. It considers both recall/sensitivity and specificity and is well-suited for imbalanced datasets.

$$recall = sensitivity = \frac{TP}{TP + FN} \tag{10}$$

$$precision = \frac{TP}{TP + FP} \tag{11}$$

$$specificity = \frac{TN}{FP + TN} \tag{12}$$

$$F1\ score = \frac{2 * precision * recall}{precision + recall} \tag{13}$$

$$balanced\ accuracy\ score = 0.5 * (sensitivity + specificity) \tag{14}$$

4 Experiments and Discussion

In this section, a number of experiments on two artificial datasets are conducted and discussed. The experiments are divided into two scenarios. Experiments under the first scenario shows how those factors including sample size, data scattering degree, and overlap degree affect classification performance of different classifiers. Experiments of the second scenario focus on how the nature of minority class such as having small disjuncts degrades classification performance.

4.1 Artificial Datasets

To explore joint effects of class imbalance and class overlap and to explore factors influencing binary classification performance, we generate two artificial datasets T_1 and T_2, as are shown in Fig. 1. In Fig. 1(a) and (b), the two plots are scattering plot and joint PDF plot, respectively. For simplicity, we denote T_P for minority class and T_N for majority class, respectively. For T_P in T_2, it is divided into three regions. From top to bottom, the first region denotes higher density of T_P, the second is small disjunct, and the third represents sparse samples which probably are noises.

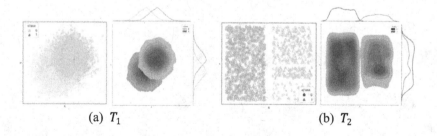

(a) T_1 (b) T_2

Fig. 1. T_1 is subject to Gaussian distribution and T_2 is subject to uniform distribution.

The distribution T_1 is subject to Gaussian distribution. To analyze the factors affecting classification performance, we control three parameters, i.e., sample size, STD, and distance between two cluster centers. Through changing the sample size and/or STD of a class, the density of the class can be adjusted more or less. Through changing the distance of two cluster centers, the degree of overlap can be controlled. The distribution T_2 is subject to Uniform distribution and only the distances between different regions from minority class and region from majority is changed to express the degree of overlapping.

In the following experiments, joint probability density plots are used widely for the aim of showing the density distribution of different regions in data space, thus bringing intuitive understanding about classification performance.

4.2   Experiment I: Factors Affecting Binary Classification Performance

In this experiment, T_1 is used to analyze factors affecting binary classification performance. Two groups of experiments are devised, namely, linearly separable circumstance and non-linearly separable one. Four IRs, which are 1.0, 2.0, 5.0, and 10.0, respectively, are chosen in the following experiments. IR, which is the abbreviation of imbalance ratio, means the ratio of number of majority samples to that of minority samples, is usually used to describe the imbalance degree of a dataset. When IR is equal to 1.0, the two classes are balanced distributed; the bigger the IR is, the more imbalanced the two classes are distributed. In addition, the ratio of the size of training set to testing set is 4:1 in all experiments.

Under the linearly separable case, the centers of the T_P and T_N are far from each other for the aim of avoiding class overlap. All metrics always get the highest result 1.0. The results prove that, under the linearly separable case, classification performance of different classifiers is not affected by imbalance ratio, dataset size, and data scattering degree.

Furthermore, regardless of other factors, as long as T_P and T_N are linearly separable, there exists a little difference among decision boundaries in terms of classifiers. The boundaries of K-NN and RF are more winding than the other four classifiers. The reasons behind that probably are that the K-NN classifier adapts more local learning strategies while the RF classifier chooses samples randomly (Fig. 2).

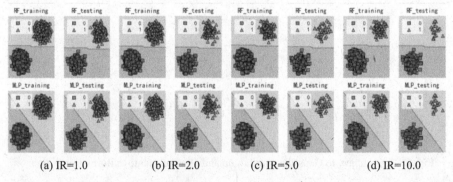

(a) IR=1.0   (b) IR=2.0   (c) IR=5.0   (d) IR=10.0

Fig. 2. Comparisons between Decision boundaries of RF classifier and MLP classifier.

But when class overlap occurs, that is, T_P and T_N become non-linearly separable, the classification results seem more complicated. In the following experiments, T_P and T_N become closer enough through changing their centers, thus leading to the occurrence of class overlap. For the convenience of discussion, the size of T_1 is fixed to 1000, and the following two circumstances are set.

Circumstance 1: T_P and T_N have same STDs but different IRs, which is shown as Fig. 3. Although they have same STDs, densities of overlapped region change with change of IRs, leading to different classification performance for different classifiers.

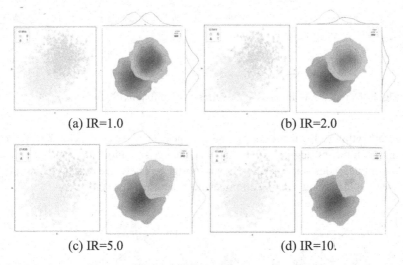

(a) IR=1.0   (b) IR=2.0

(c) IR=5.0   (d) IR=10.

Fig. 3. Two classes with same STDs but different IRs in non-linearly separable classification.

Generally speaking, when the IR is higher, both T_P and overlapped region are distributed more sparsely, but the density of T_N is the highest and the overlapped regions are medium, which leading to the classifier bias T_N seriously. Just as is shown in Fig. 4, when the IR = 10.0, recall values of all the six classifiers are decreased greatly.

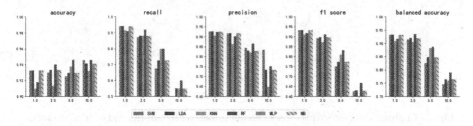

Fig. 4. Comparisons on performance metrics under different IRs for testing dataset.

Comparisons in Fig. 5 among different classifiers show that the KNN classifier performs worst among six classifiers while the NB classifier performs best. The underlying reason may be the densities of samples from T_P and T_N are both higher and same, therefore, compared with the other five global learning algorithms, for the KNN classifier, which uses a local learning algorithm, is more difficult to recognize samples from T_N and T_P correctly. As for the NB classifier, it is proved that when its conditional independence is held, it always obtains very remarkable classification performance.

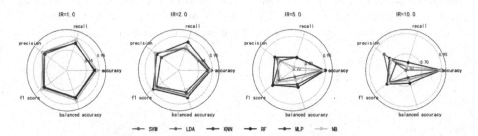

Fig. 5. Comparisons among six classifiers under different IRs for testing dataset.

Keep everything else the same, if the distance between centers of T_P and T_N is changed, which means degrees of overlap change, classification performance of all the six classifiers degrades seriously, as shown in Fig. 6. Therefore, class overlap plays a critical role in classifying samples in imbalanced datasets correctly.

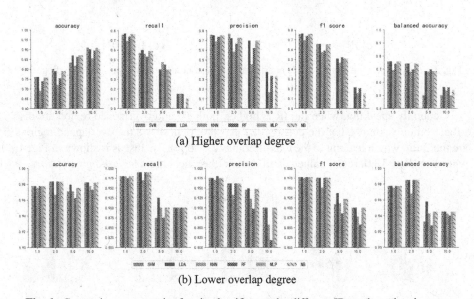

(a) Higher overlap degree

(b) Lower overlap degree

Fig. 6. Comparisons on metrics for six classifiers under different IRs and overlap degrees.

Circumstance 2: T_P and T_N have different STDs and IRs. Figure 7 is an example which only shows two balanced cases with different STDs. Obviously, if T_P and T_N are balanced, compared with Fig. 3(a), the densities around two centers, overlapped regions, and other regions seem clearly different when they have different STDs. The differences of densities make difference in classification performance. In fact, the similar results can be obtained from imbalanced classes with different stand deviations.

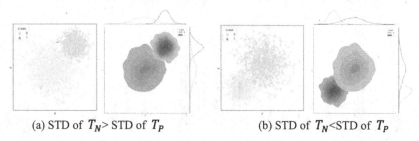

(a) STD of T_N> STD of T_P (b) STD of T_N<STD of T_P

Fig. 7. Two balanced classes with different STDs in non-linearly separable classification.

According to the experimental results on testing datasets, some conclusions can be made as follows.

Firstly, all metrics under the case of T_P and T_N with different STDs are higher than the case where they have same STDs. The reason on the one hand is that the density of T_N is no more dominant, and the classifiers do not bias it so seriously. The reason on the other hand is, when IR is very high, the whole data space become sparser, and it benefits the recognition of T_P and T_N simultaneously.

Secondly, when T_P has smaller STD and the IR is not so high, its density will be higher, which benefits the correct recognition of T_P samples, and thus improving the F1 scores. But when the IR is very high, although the recall is a little lower, the higher density of T_N makes the classifier bias it, and thus leading to the metrics on the T_P and T_P with small STD are lower than that with big STD.

Figure 8 shows the comparisons among different IRs on metrics for T_P and T_N with different STDs in testing dataset. In terms of classifiers, comparatively, classifier RF and classifier KNN are better, and the classifier NB is still the best, as shown in Fig. 9.

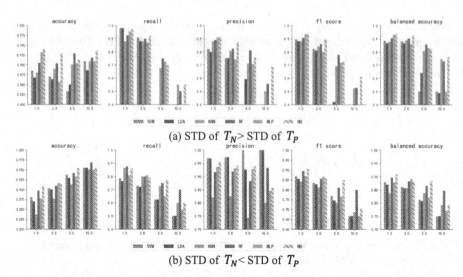

(a) STD of $T_N >$ STD of T_P

(b) STD of $T_N <$ STD of T_P

Fig. 8. Comparisons on metrics about IRs and classifiers when T_P and T_N have different STDs.

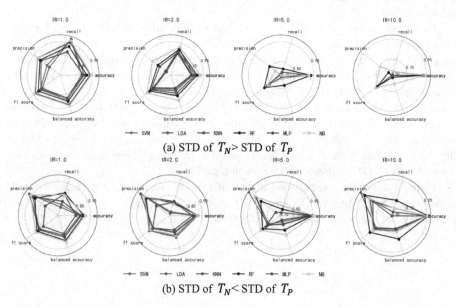

(a) STD of $T_N >$ STD of T_P

(b) STD of $T_N <$ STD of T_P

Fig. 9. Comparisons on metrics among six classifiers when T_P and T_N have different STDs.

Similarly, through changing the distance of two centers, the degree of overlap is changed accordingly. Figure 10 shows comparisons on metrics in terms of IRs and classifiers when overlap degree is different.

Because the real-world data is larger, keep other parameters the same and only the dataset is expanded to 10000, we found that the general trend in all metrics is decreasing more or less with the increasing of dataset under balanced case. The decreasing trend

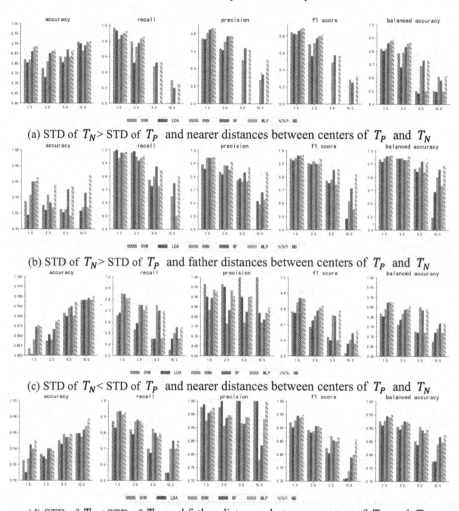

(a) STD of T_N > STD of T_P and nearer distances between centers of T_P and T_N

(b) STD of T_N > STD of T_P and father distances between centers of T_P and T_N

(c) STD of T_N < STD of T_P and nearer distances between centers of T_P and T_N

(d) STD of T_N < STD of T_P and father distances between centers of T_P and T_N

Fig. 10. Comparisons on metrics about IRs and classifiers when overlap degree is different.

indicates that the two balanced classes regardless of STDs are more difficult to be separated and the reason is that the increase on number of samples make each part of data space have similar higher density. Furthermore, if there is difference between STDs, the influence brought by changes of STD on classification performance keep consistent with previous case where the dataset is only 1000.

Among six classifiers, most classifiers show similar classification performance to previous cases and classifier NB is still the most robust and perform best, as shown in Fig. 11.

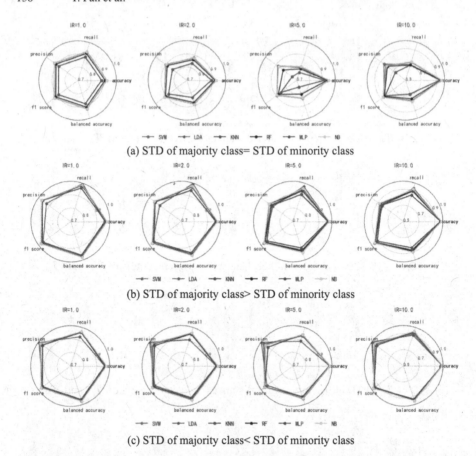

(a) STD of majority class= STD of minority class

(b) STD of majority class> STD of minority class

(c) STD of majority class< STD of minority class

Fig. 11. Comparisons among six classifiers when T_P and T_N have different STDs and IRs.

4.3   Experiment II: Influences of Small Disjuncts and Noise

We now study how different regions of T_P from T_2 affect the classification performance when class imbalance and class overlap occur simultaneously. The number of samples in T_N is 500. The number of three different regions of T_P is 300, 150 and 50. Figure 12 shows a case where high density of T_P is overlapped with T_N. We change the overlapped degrees through adjusting distances between T_N and different regions of T_P. The results show both small disjuncts and noises degrade the classification performance, comparatively speaking, small disjuncts aggravate the classification performance more than noises.

When the size of the dataset is not so big, just like our artificial dataset whose size is only 1500, classifier MLP and classifier NB perform better than other classifiers, as shown in Fig. 12.

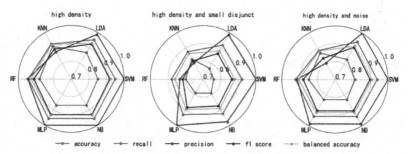

Fig. 12. Metrics comparisons among different classifiers when different regions of T_P are overlapped with T_N.

5 Conclusions and Future Work

In this study, we describe an experimental study aimed at identifying the joint effects of class imbalance and class overlap. Two artificial datasets subject to Gaussian distribution and uniform distribution are used to control factors affecting classification performance easily. The following four conclusions can be made according to our plentiful experiments. Firstly, under the linearly separable case, IRs and factors like STD for Gaussian distribution have no influence on classification performance. Secondly, under the non-linearly separable case, factors like class overlap, size of dataset and STDs for Gaussian distribution will bring different changes to the density of different regions in data space, and thus affecting the binary classification performance. Thirdly, regions with different densities from minority class, such as small disjuncts and noise, degrade the classification to different extent. And lastly, based on the two artificial datasets used in this work, classifier NB is the best and classifier KNN performs worst. According to the above conclusions, to decrease the joint effects of class imbalance and class overlap, it is feasible to change densities of samples from two classes. Meanwhile, more robust classifier should be devised specially to fit to imbalanced datasets even with class overlap.

The focus of this work is on the binary classification, further research is necessary to do in multiple classification scenario. Furthermore, it is clearly not sufficient to only use simple two-dimensional artificial datasets to analyze joint effects of class imbalance and class overlap, real-world datasets with larger size and high dimensions should be adopted in the future. In addition, with the popular development of deep learning, more and more outstanding classifiers, such as CNN and GAN, etc. are widely used and show decent classification performance. Therefore, validation for these deep learning algorithms also should be done to make sure whether deep learning algorithms help address both class imbalance and class overlap.

Acknowledgments. The authors acknowledge National Natural Science Foundation of China (Grant: 62066039), Natural Science Foundation of Qinghai Province (Grant: 2022-ZJ-925), and the "111" Project (D20035).

References

1. Shaukat, A.S., Usha, A.: An effective distance-based feature selection approach for imbalanced data. Appl. Intell. **50**, 717–745 (2020)
2. Dai, Q., Liu, J.W., Shi, Y.H.: Class-overlap undersampling based on Schur decomposition for Class-imbalance problems. Expert Syst. Appl. **221**, 119735 (2023)
3. Hoyos-Osorio, J., Alvarez-Meza, A., et al.: Relevant information undersampling to support imbalanced data classification. Neurocomputing **436**, 136–146 (2021)
4. Li, D.-C., Wang, S.-Y., et al.: Learning class-imbalanced data with region-impurity synthetic minority oversampling technique. Inf. Sci. **607**, 1391–1407 (2022)
5. Vuttipittayamongkol, P., Elyan, E., Petrovski, A.: On the class overlap problem in imbalanced data classification. Knowl.-Based Syst. **212**, 106631(2021)
6. Lee, H.K., Kim, S.B.: An overlap-sensitive margin classifier for imbalanced and overlapping data. Expert Syst. Appl. **98**, 72–83 (2018)
7. Barella, V.H., Garcia, L.P.: Assessing the data complexity of imbalanced datasets, Inf. Sci. **553**, 83–109 (2021)
8. Dudjak, M., Martinović, G.: An empirical study of data intrinsic characteristics that make learning fromimbalanced data difficult. Expert Syst. with Appl. **182** (2021)
9. Santos, M.S., Abreu, P., et al.: A unifying view of class overlap and imbalance: key concepts, multi-view panorama, and open avenues for research. Inf. Fus. **89**, 228–253 (2023)
10. IBM homepage. https://www.ibm.com/topics/naive-bayes
11. García, V., Sánchez, J., Mollineda, R.An empirical study of the behavior of classifiers on imbalanced and overlapped datasets. In: Progress in Pattern Recognition, Image Analysis and Applications, 12th Iberoamericann Congress on Pattern Recognition, CIARP 2007, Valparaiso, Chile, pp. 397–406(2007)
12. García, V., Mollineda, R.A., Sánchez, J.S.: On the k-NN performance in a challenging scenario of imbalance and overlapping. Pattern Anal. Appl. **11**(3), 269–280(2008)
13. Lee, H.K., Kim, S.B.: An overlap-sensitive margin classifier for imbalanced and overlapping data. Expert Syst. Appl.**98**, 72–83(2018)
14. Linear Discriminant Analysis. https://www.geeksforgeeks.org/
15. Bishop, C.: Neural Networks for Pattern Recognition. Oxford University Press, USA (1995)
16. Yuan, B.W., Zhang, Z.L., et al.: OIS-RF: a novel overlap and imbalance sensitive random forest. Eng. Appl. Artif. Intell. **104**, 104355 (2021)
17. Duda, R.O., Hart, P.E., Stork, D.G.: Pattern Classification and Scene Analysis. Wiley, New York (2001)
18. Liang, X.W., Jiang, A.P., et al.: LR-SMOTE—An improved unbalanced dataset oversampling based on K-means and SVM. Knowl.-Based Syst. **196**, 105845 (2020)
19. Shi, S., Li, J., et al.: A hybrid imbalanced classification model based on data density. Inf. Sci. **624**, 50–67 (2023)
20. Wei, Z., Zhang, L., Zhao, L.: Minority-prediction-probability-based oversampling techniquefor imbalanced learning. **622**, 1273–1295 (2023)
21. Han, H., Li, W., Wang, J., Qin, G., Qin, X.: Enhance explainability of manifold learning. Neurocomputing **500**, 877–895 (2022). https://doi.org/10.1016/j.neucom.2022.05.119

Evaluation of SVM Transformations for Multi-label Research Article Classification

Amr S. Abdelfattah[1] and Tomas Cerny[2]

[1] Baylor University, One Bear Place #97141, Waco, TX, USA
amr_elsayed1@baylor.edu
[2] SIE, University of Arizona, Tucson, AZ, USA
tcerny@arizona.edu

Abstract. Support Vector Machine (SVM) models are often used to address the problem of multi-labeled topic classification across various datasets. However, to fit the SVM-required modeling, it is necessary to transform the multi-labeled dataset into single-labeled ones, which plays a crucial role in the performance. Different transformations applied to the problem may lead to varying accuracy and behavior of SVM over datasets. This paper investigates the performance of the SVM model on a multi-labeled research articles dataset, with three commonly used transformations: Binary Relevance (BR), Label Powerset, and Classifier Chains. It also performs a case study of categorizing research articles into one or more related topics. The results indicate that the Label Powerset transformation achieved the best average accuracy score across all topics classification. Moreover, the Label Powerset and BR transformations were able to achieve a Hamming loss measurement of $\approx 90\%$ for the fraction of topics that are incorrectly assigned. However, BR exhibited the best recall and precision balancing for class classification measurements. The paper introduces the Least Class Classifier (LCC) technique, which challenges the problem of imbalanced datasets to achieve an equal chance for the minor classes. This technique addresses the problem of imbalanced datasets. It also shows promising results for increasing recall calculations for the minor class in imbalanced datasets. That emphasizes the performance of SVM models for multi-labeled topic classification can be improved by selecting an appropriate transformation technique.

Keywords: Support Vector Machine · SVM · Multi-labeled Topic Classification · Multi-class classification · Binary Relevance · Classifier Chains · Label Powerset

1 Introduction

Topic classification is a subfield of topic analysis that involves identifying the topics of a set of texts before analyzing them. Depending on the nature of the dataset, topic classification can be of different types, including binary class and multi-class. This paper focuses on multi-labeled topic classification, which is

© The Author(s), under exclusive license to Springer Nature Switzerland AG 2024
H. Han and E. Baker (Eds.): SDSC 2023, CCIS 2113, pp. 141–157, 2024.
https://doi.org/10.1007/978-3-031-61816-1_10

performed on an articles dataset to assign an article to one or more classes [2]. A topic represents a theme or an underlying idea presented in a text. By learning these topics, a topic classifier can make predictions on new texts [3,12].

Text classification tasks primarily involve finding the best combination of features and machine learning models for the problem and dataset. Among the many classification methods, SVM has demonstrated superior performance [4]. However, SVM was initially developed for binary decision problems, and extending it to multi-class problems is challenging from the perspective of varying performance. The main focus of this paper is to investigate the impact of different transformation techniques on the measurements of the SVM model in multi-labeled classification. The paper examines the performance of the SVM model on the following three transformation techniques: Binary Relevance, Classifier Chains, and Label Powerset. They are applied to a multi-labeled articles dataset consisting of scientific paper titles and abstracts labeled through six topics. This dataset is publicly available at this link[1]. In addition, the paper proposes the Least Class Classifier (LCC) technique to support the minor classes in the imbalanced dataset, providing better chances for a high recall [2].

The organization of this paper is as follows: Sect. 2 provides the background information, while Sect. 3 presents the related work. The proposed LCC technique is explained in Sect. 4, and Sect. 5 describes the experiment and its results. Finally, Sect. 6 concludes the paper and outlines future work.

2   Background

Transformations are required to apply Support Vector Machines (SVMs) to multi-class classification problems. The common approach is to decompose multi-class problems into many binary-class problems and incorporate many binary-class SVMs [4]. However, applying SVMs to multi-labeled classification problems is more challenging due to the required transformations and extensive time complexity.

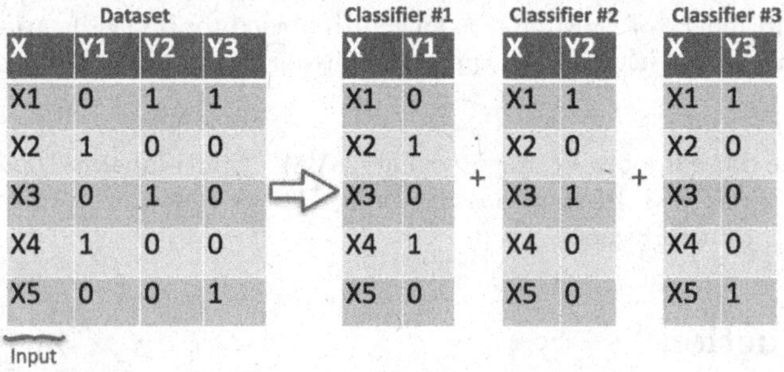

Fig. 1. Binary Relevance (BR) Example

---

[1] Research articles Dataset: https://zenodo.org/record/7671877#.Y_fFEy2B3RY, accessed on 02/23/2023.

One of the most widely used transformations for multi-labeled classification tasks is Binary Relevance (BR) [22,24]. BR transforms the task into many independent binary classification problems, as depicted in Fig. 1. The dataset is split into three classifiers, duplicating the input features (X column) with each target class (Y1, Y2, Y3 Columns) individually. BR aims to optimize the Hamming Loss and requires only one-step learning. However, it may have a class-imbalance issue and does not consider label correlations, as SVM classifiers trained on imbalanced datasets can produce suboptimal models biased towards the majority class and perform poorly on the minority class [1].

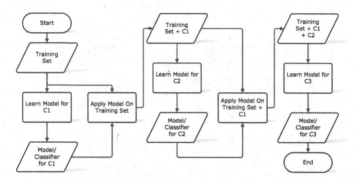

Fig. 2. The Classifier Chains Flow at [11]

To address label correlations, some practical work in [21] has highlighted two additional transformations. First, the Classifier Chains [13,21] considers the correlation by starting with a classifier trained only on the input data. Each subsequent classifier is then trained on the input space and all previous classifiers in the chain, as shown in Fig. 2. As illustrated in Fig. 3, the first classifier (Classifier #1) is similar to the first one in the BR transformer as well, while the second classifier (Classifier #2) differs by including one of the target classes (Y1) among its input to train the model accordingly. This process continues,

Dataset				Classifier #1			Classifier #2		
X	Y1	Y2		X	Y1		X	Y1	Y2
X1	0	1		X1	0		X1	0	1
X2	1	0		X2	1		X2	1	0
X3	0	1	⇨	X3	0	+	X3	0	1
X4	1	0		X4	1		X4	1	0
X5	0	0		X5	0		X5	0	0
Input				Input			Input		

Fig. 3. Classifier Chains Example

and the final output is a set of classifiers with each considering the correlation among target classes.

The Label Powerset [14,21] is the second transformation that addresses label correlations. It transforms the multi-labeled classification problem into a multi-class problem where one multi-class classifier is trained on all unique label combinations found in the training data. As illustrated in Fig. 4, one classifier is generated while producing multi-class labels from the multi-labeled target classes. For instance, the second and fourth rows are assigned to the same class because they belong to the same multi-labeled target classes.

Dataset					Classifier #1	
X	Y1	Y2	Y3		X	Y1
X1	0	1	1		X1	1
X2	1	0	0		X2	2
X3	0	1	0	⇨	X3	3
X4	1	0	0		X4	2
X5	0	1	0		X5	3

Input

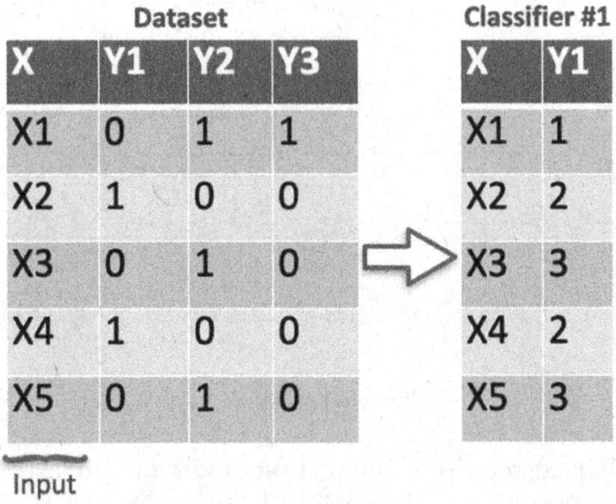

Fig. 4. Label Powerset Example

3   Related Work

Text classification has been extensively studied in the literature [10,16]. One of the earliest prominent works in this field is Joachims (1998) [8], which provides both theoretical and empirical evidence that Support Vector Machines (SVMs) are very well suited for text categorization. Joachims compared SVMs with four other approaches, including Bayes [23], Rocchio [23], C4.5 [17], and k-NN [23], and found that SVMs achieved the highest accuracy of 86.4% on two corpora, with k-NN coming in second at about 82.3% accuracy.

For multi-labeled topic classification, Sebastiani [19] stated that an algorithm for binary classification can also be used for multi-labeled topic classification by transforming the problem of multilabel classification into independent problems of binary classification. However, Sebastiani emphasized that this approach requires categories to be stochastically independent of each other. The paper also compared SVM with Decision Tree (DT), obtaining an accuracy of 92% for SVM and 88.4% for DT using the Reuters corpus [18].

More researches [6,7] balances the effect of skewed training dataset by the Different Error Costs method (DEC) and constructs a predictor using the support vector machine as a classifier for a corresponding dataset. However, the impact of the applied transformation on the SVM model is not highlighted and it could impact the accuracy of the model.

Sun et al. [20] used the BR transformation strategy to realize multi-label classification effectively. They applied the divide-and-conquer strategy to divide the representative set into subsets, ensuring that each representative subset contains a certain number of positive and negative instances. They also applied the DEC method to overcome the label imbalance problem.

Wu et al. [22] presented a novel multi-label classification model that combines Rank-SVM and BR with robust Low-rank learning (RBRL). RBRL inherits the ranking loss minimization advantages of Rank-SVM, overcoming the disadvantages of BR, which suffers from the class imbalance issue and ignores label correlations.

This paper investigates the three transformations mentioned above (Binary Relevance, Classifier Chains, and Label Powerset) and their impact. It also proposes and evaluates an LCC technique for considering the least presenting classes in imbalanced datasets.

4   The Proposed Least Class Classifier (LCC)

This section details the LCC technique proposed to tackle the impact of imbalanced datasets on classification problems, with the objective of increasing the recall of the minor class in such datasets. The proposed technique is outlined as follows:

1. Multiple binary classifiers are built between the minor class and all other classes, one at a time, resulting in K-1 binary classifiers for K classes.
2. The dataset is adjusted for each classifier by removing the common and non-present articles between the two classes. For instance, when building a model for classifying the Computer Science and Quantitative Finance topics, the articles that are common or not presented in both topics are eliminated.
3. Each classifier is applied, and only the predicted positive class values, which are the articles classified as positive for this minor class, are considered. This process aims to increase the chance for this minor class to be identified with each other class separately.
4. The positive class predictions from all k-1 classifiers are combined to construct the positive predictions for this minor class. Any article classified as positive for this minor class by any of these classifiers is considered as classified for this class. After that, to produce a complete class prediction, these results are merged with the results of a complete transformation classifier (i.e., BR, Classifier Chains, Label Power) over all classes.

This technique intends to balance the impact of imbalanced datasets on classification problems by considering the least presented classes. By doing so, it provides an equal opportunity for all classes to be identified regardless of their balance in the dataset.

5   Experiment

In this section, we detail the experimental procedure that follows the depicted process in Fig. 5. The study starts with a dataset of research articles and applies pre-processing and vectorization to convert the abstract and title into a feature vector. The transformation selection stage applies four transformers (Binary Relevance, Classifier Chains, Label Powerset, and Least Class Classifier) one at a time. Then, the SVM training model is built using balancing weights and cross-validation. Finally, the testing process applies the trained SVM model to new data to classify articles into their corresponding topics. The experimental results are evaluated using confusion matrices, accuracy scores, precision, recall, F1 score, and Hamming loss rank. It shows a results comparison among the four transformation techniques. The following sections draw the details of these phases' execution.

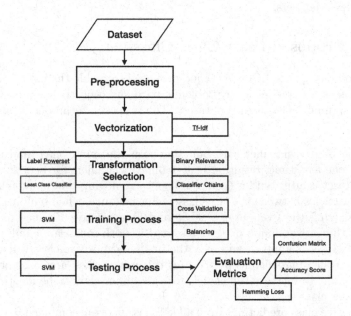

Fig. 5. The methodology phases

5.1   Dataset Description

In this section, we discuss the dataset (See footnote 1) used for the study, which consists of scientific articles with three columns: ID, Title, and Abstract. These articles are classified into six different topics, namely Computer Science (CS), Physics (Phys), Mathematics (Math), Statistics (Stats), Quantitative Biology (QB), and Quantitative Finance (QF). The dataset is analyzed to reveal that 66% of articles are assigned to a single topic, and only three topics are marked per article as the maximum, as shown in Table 1. Additionally, Table 2 presents the number of articles per topic and each pair of topics at the same time. The analysis demonstrates that Computer Science, Physics, and Quantitative Biology are the most commonly assigned individual classes, while Computer Science and Statistics are the most commonly paired classes.

The dataset used for this study is a downsampled version of the one provided in [9]. The original dataset includes 20,972 articles categorized into six topics as illustrated in Table 3. However, this dataset suffers from a significant imbalance, particularly with the underrepresented Quantitative Biology and Quantitative Finance topics. We performed down-sampling as follows: we kept the two minor classes with their original number of articles, and we randomly selected 10% of articles from each of the other topics. This approach keeps 3309 articles. Although it sacrifices some real data but helps to mitigate the imbalance between classes and reduce the required computation time for training the models, as shown in Table 4. It is worth noting that the LCC technique will be applied to the complete dataset, as explained later.

The process of splitting the dataset into distinct portions is an important aspect of the experimental study. Specifically, the dataset is divided into two parts: 70% is allocated for the Training Phase and the remaining 30% for the Testing Phase. This ensures that the trained model is tested on a separate set of data that was not used for training, in order to accurately assess its performance on unseen data. This approach also reduces the risk of overfitting the model to the training data, which could result in poor generalization to new data. The splitting process is done randomly, and the resulting subsets are representative of the dataset in terms of class distribution.

Table 1. The Number of Topics assigned to articles at the same time

Number Classes	# Items
One Class	2178 (66%)
Two Classes	1059 (32%)
Three Classes	72 (2%)
All	3309

Table 2. Number of Articles that have assigned to each Pair of topics *Number in parentheses = count of sole-topic articles.*

Topics	CS	Stats	Math	Phys	QB	QF
Computer Science (CS)	(493)	513	187	93	35	11
Statistics (Stats)	513	(150)	206	30	110	27
Mathematics (Math)	187	206	(367)	58	0	1
Physics (Phys)	93	30	58	(516)	0	0
Quantitative Biology (QB)	35	110	0	0	(443)	4
Quantitative Finance (QF)	11	27	1	0	4	(209)

Table 3. The Original Dataset

Topics	# Items
Computer Science	8594 (41%)
Physics	6013 (29%)
Mathematics	5618 (27%)
Statistics	5206 (25%)
Quantitative Biology	587 (3%)
Quantitative Finance	249 (1%)
All	20972

Table 4. The Down-sampled Dataset

Topics	# Articles
Computer Science	1261 (38%)
Statistics	968 (29%)
Mathematics	764 (23%)
Physics	683 (21%)
Quantitative Biology	587 (17%)
Quantitative Finance	249 (7%)
All	3309

5.2 Pre-processing

The pre-processing phase is a set of steps that aim to generate a vector of features from the input text of the articles. Firstly, the title and abstract are weighted equally by concatenating them to represent the article input text as a single value. Then, all non-alphabetical characters and stop words (e.g., and, or, is) are removed. The article text is tokenized, such that each word is considered a token. Finally, the words' tokens are converted into their lemmas. Lemmatization is the process of transforming a word to its base or root form, which helps to reduce the complexity of the text and improve the accuracy of natural language processing tasks. For example, the word "players" is converted to "player", and "went" is converted to "go". This approach ensures a consistent methodology for considering the semantics of words. After applying this phase to the dataset, Table 5 shows the number of tokens and lemmas extracted from the article data.

Table 5. The Dataset Tokens and Lemmas

Topics	# Items	# Tokens	# Lemmas
All	3309	598040	347162

5.3 Vectorization

The vectorization phase is an important step that transforms text into a meaningful vector of numbers that a machine learning model can understand. This experiment employs the *Term Frequency * Inverse Document Frequency (Tf-Idf)* [12]. This method takes into account how common a word is in all articles versus how common it is in a specific article. Therefore, more common words across the articles are ranked higher since they are considered a better representation of a document. The Tf-Idf technique is used on the articles in the dataset, resulting in an average feature vector length of 15,836. The vectorization process maps each article to a vector of numbers, where each number represents the importance of a particular term in the article. This is achieved by assigning a weight to each term in the article based on its frequency in the article and across all articles. The result is a numerical representation of the article that can be used as input for machine learning algorithms.

5.4 Training Process

The SVM model is trained on the aforementioned training dataset. The balanced weight technique is applied, which is one of the widely used methods for imbalanced classification models. It modifies the class weights of the majority and minority classes to reduce bias towards the majority class in the dataset.

Cross-validation [15] is used to determine the best fitting combinations among Linear, Sigmoid, Radial Basis Function (RBF), and second-degree Polynomial kernels with different values for regularization factor and kernel coefficient. The 5-Fold cross-validation method is employed, which involves dividing the training data into five equal parts. Out of these, four parts are used for training, while the remaining part is reserved for validation. This process is repeated five times, with each part serving as the validation set once. The Hamming Loss [5] is chosen as a metric to evaluate the accuracy of different parameters. It measures the fraction of labels that are incorrectly predicted, which is particularly relevant for multi-label classification tasks. The chosen cross-validation parameters for the three transformations are summarized in Table 6. The LCC in this study uses the BR transformation when it merges the results, therefore, it utilizes those same parameters as well.

5.5 Testing Process

In the testing phase, the SVM model performance is evaluated and compared for the three transformation methodologies: Binary Relevance, Classifier Chains, and Label Powerset. The Hamming loss is used to evaluate the accuracy of the predicted labels, where the fraction of incorrectly predicted labels is calculated. Additionally, the accuracy score is calculated to indicate the accuracy of the set of labels predicted for a sample, which must exactly match the corresponding set of topics in the dataset. Furthermore, the per-class accuracy is illustrated through the precision and recall measurements for each transformation in Tables

7, 8, and 9. This allows for highlighting the performance of each separate topic. The confusion matrices in Figs. 6, 7, and 8 summarize the performance of the classification algorithm for each transformation.

For interpretation of the confusion matrices, it is worth mentioning that they provide a visual representation of the performance of the classification algorithm for each transformation. They are used to illustrate the classification errors and correct classifications made by the SVM model. The rows of the matrix represent the true labels, while the columns represent the predicted labels. The diagonal entries of the matrix represent the number of correctly classified instances, while the off-diagonal entries represent the incorrectly classified instances. The values in the confusion matrices are the number of instances that belong to each category, which are classified into each category. Therefore, by analyzing the confusion matrices, the performance of each transformation can be compared to the other transformations, as well as identifying the topics that the SVM model struggles to classify accurately.

5.6   Addressing the Imbalanced Dataset Using LCC

The main objective of the LCC technique is to tackle the issue of an imbalanced dataset for the least presented classes. The original dataset (before the down-sampling process) discussed in this study exhibited significant data imbalance, as demonstrated in Table 3. The data revealed that the Quantitative Finance topic is the least frequent class, representing only 1% of the dataset. Therefore, the LCC technique is applied to the original dataset [9].

Table 6. The Cross Validation Parameters Linear $= \langle x, x' \rangle$ and RBF $= \exp(\gamma \|x - x'\|^2)$; γ is Kernel Coefficient

Transformations	Kernel	Regularization Factor	Kernel Coefficient
Binary Relevance	Linear	1.0	–
Classifier Chains	RBF	1.0	1.0
Label Powerset	Linear	1000	–

To execute the LCC, apply the steps described in Sect. 4. The results obtained for the Quantitative Finance class are compared with the outcome of the BR transformation. The BR transformation demonstrates a low recall measurement, indicating that it is unlikely to contain true positive articles classification as shown in Fig. 9a. However, LCC succeeds in achieving a higher recall for the

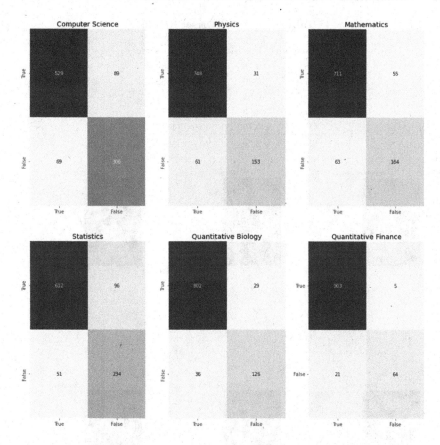

Fig. 6. The Binary Relevance Confusion Matrix

positive class, as shown in Fig. 9b. Although the precision is lower compared to the BR transformation, the LCC combines its predicted positive class results only with the complete prediction results from another classifier (e.g., BR), thus achieving the produced high recall with better precision inherited from the other integrated classifier.

Table 7. Binary Relevance Results

Categories	Precision	Recall	F1-score
CS	0.77	0.82	0.79
Phys	0.83	0.71	0.77
Math	0.75	0.72	0.74
Stats	0.71	0.82	0.76
QB	0.81	0.78	0.79
QF	0.93	0.75	0.83

Table 8. Classifier Chains Results

Categories	Precision	Recall	F1-score
CS	0.78	0.76	0.77
Phys	0.85	0.68	0.75
Math	0.79	0.67	0.72
Stats	0.72	0.71	0.72
QB	0.64	0.75	0.69
QF	0.79	0.78	0.78

Table 9. Label Powerset Results

Categories	Precision	Recall	F1-score
CS	0.78	0.81	0.80
Phys	0.90	0.74	0.82
Math	0.81	0.66	0.72
Stats	0.75	0.67	0.71
QB	0.75	0.80	0.77
QF	0.89	0.76	0.82

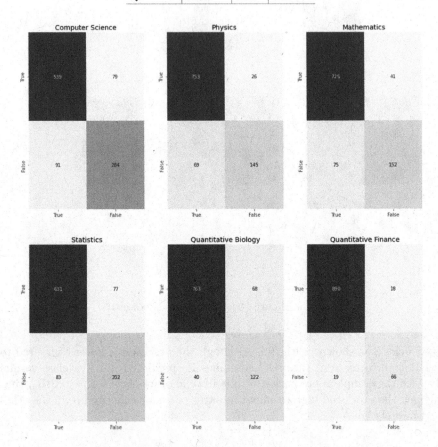

Fig. 7. The Classifier Chains Confusion Matrix

5.7   Results Analysis

The results, as presented in Fig. 10, indicate that Label Powerset achieved the highest accuracy score among the three methods, which can be attributed to its ability to optimize the classification performance based on all assigned classes as

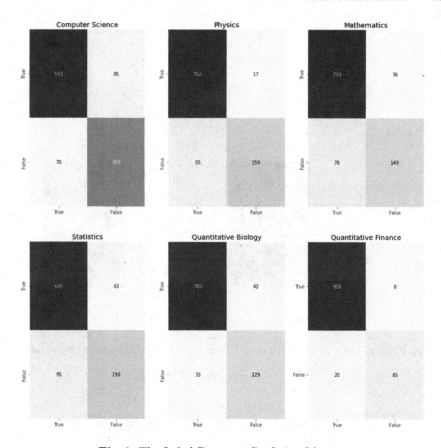

Fig. 8. The Label Powerset Confusion Matrix

a whole. However, BR and Label Powerset had similar Hamming loss measurements, which were the best among the three methods at approximately 90%. In contrast, Classifier Chains had slightly lower accuracy and Hamming loss scores than the other methods. The precision and recall measurements analysis revealed that BR had the best balance between both measurements, particularly in terms of achieving higher recall per each class separately. Although Classifier Chains transformation achieved a balance between the two measurements, it had lower scores, particularly for the minor classes. Label Powerset had better precision than recall.

On the other hand, the original dataset before downsampling showed a significant data imbalance, particularly for the Quantitative Finance topic, which had only 1% of the total dataset. The LCC technique shows advancement in

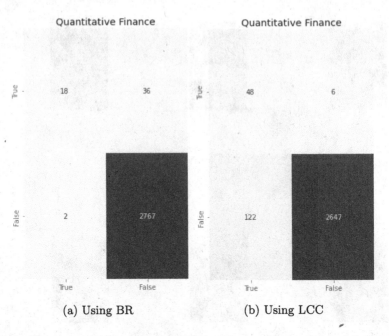

(a) Using BR   (b) Using LCC

Fig. 9. Classifying the Quantitative Finance Topic over the Original Dataset

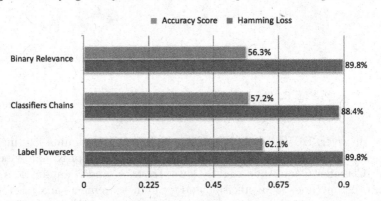

Fig. 10. Transformations' Results comparison

addressing this problem of handling underrepresented classes. The results in Fig. 11 demonstrate that the LCC technique improved the recall for the minority class while maintaining a satisfactory level of precision. In comparison to solely the BR method, which had a low recall score of 33% for the Quantitative Finance topic, the LCC method achieved a promising recall measurement of 89%.

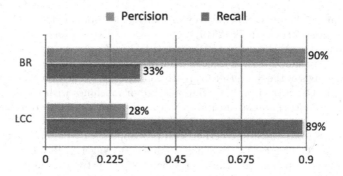

Fig. 11. BR and LCC Results comparison

6   Conclusion

In this study, the BR, classifier chains, and Label Powerset transformations were compared in terms of their accuracy for multi-label classification using the SVM model. Each transformation had a different behavior and requires tailored setup parameters that were addressed in the training phase. While Label Powerset produced the best overall accuracy score and Hamming loss measurements for all topics, BR was the best method for balancing precision and recall, particularly for each class separately. Classifier Chains transformation was not the best method for data imbalances, especially for the minor classes.

The results showed that data imbalance had a significant impact on measurements and classification, particularly for the minor classes. The LCC technique was introduced as a promising solution for imbalanced datasets, especially for underrepresented classes. By integrating the LCC with an additional classifier, it achieved higher recall for the minority class while maintaining a satisfactory level of precision.

Future work could involve the examination of the behavior of the proposed approach on various datasets. In addition, it would be worthwhile to investigate the effectiveness of the classifier chains transformation with different class orders, in order to ascertain if it leads to an improvement in accuracy. Furthermore, there is a need for more research in the pre-processing phase to determine the impact of dimensionality reduction and independence measurements on the features.

Acknowledgement. This material is based upon work supported by the National Science Foundation under Grant No. 1854049 and a grant from Red Hat Research. I would like to express my sincere gratitude to Dr. Greg Hamerly for his invaluable guidance and expertise.

References

1. Batuwita, R., Palade, V.: Class imbalance learning methods for support vector machines. In: Imbalanced Learning: Foundations, Algorithms, and Applications, pp. 83–99 (2013)

2. Burkhardt, S., Kramer, S.: A survey of multi-label topic models. ACM SIGKDD Explor. Newsl. **21**(2), 61–79 (2019)
3. Chen, Q., Allot, A., Leaman, R., Doğan, R.I., Lu, Z.: Overview of the biocreative vii litcovid track: multi-label topic classification for covid-19 literature annotation. In: Proceedings of the Seventh BioCreative Challenge Evaluation Workshop (2021)
4. Cheong, S., Oh, S.H., Lee, S.Y.: Support vector machines with binary tree architecture for multi-class classification. Neural Inf. Process.-Lett. Rev. **2**(3), 47–51 (2004)
5. Destercke, S.: Multilabel prediction with probability sets: the hamming loss case. In: Laurent, A., Strauss, O., Bouchon-Meunier, B., Yager, R.R. (eds.) IPMU 2014. CCIS, vol. 443, pp. 496–505. Springer, Cham (2014). https://doi.org/10.1007/978-3-319-08855-6_50
6. Hasan, M.A.M., Ahmad, S.: mLysPTMpred: multiple lysine PTM site prediction using combination of SVM with resolving data imbalance issue. Nat. Sci. **10**(9), 370–384 (2018)
7. Hasan, M.A.M., Li, J., Ahmad, S., Molla, M.K.I.: predCar-site: carbonylation sites prediction in proteins using support vector machine with resolving data imbalanced issue. Anal. Biochem. **525**, 107–113 (2017)
8. Joachims, T.: Text categorization with support vector machines: learning with many relevant features. In: Nédellec, C., Rouveirol, C. (eds.) ECML 1998. LNCS, vol. 1398, pp. 137–142. Springer, Heidelberg (1998). https://doi.org/10.1007/BFb0026683
9. Kaggle: Topic modeling for research articles (2020). https://www.kaggle.com/blessondensil294/topic-modeling-for-research-articles. Accessed 18 Aug 2020
10. Khanal, B., Rivas, P., Orduz, J.: Human activity classification using basic machine learning models. In: 2021 International Conference on Computational Science and Computational Intelligence (CSCI), pp. 121–126. IEEE (2021)
11. Nooney, K.: Deep dive into multi-label classification with detailed case study (2019). https://towardsdatascience.com/journey-to-the-center-of-multi-label-classification-384c40229bff
12. Rahman, M.A., Akter, Y.A.: Topic classification from text using decision tree, k-NN and multinomial Naïve Bayes. In: 2019 1st International Conference on Advances in Science, Engineering and Robotics Technology (ICASERT), pp. 1–4. IEEE (2019)
13. Read, J., Pfahringer, B., Holmes, G., Frank, E.: Classifier chains: a review and perspectives. J. Artif. Intell. Res. **70**, 683–718 (2021)
14. Read, J., Puurula, A., Bifet, A.: Multi-label classification with meta-labels. In: 2014 IEEE International Conference on Data Mining, pp. 941–946. IEEE (2014)
15. Refaeilzadeh, P., Tang, L., Liu, H.: Cross-validation. In: Liu, L., Özsu, M.T. (eds.) Encyclopedia of Database Systems, vol. 5, pp. 532–538. Springer, Boston (2009). https://doi.org/10.1007/978-0-387-39940-9_565
16. Sooksatra, K., Khanal, B., Rivas, P.: On adversarial examples for text classification by perturbing latent representations (2024)
17. Ruggieri, S.: Efficient c4. 5 [classification algorithm]. IEEE Trans. Knowl. Data Eng. **14**(2), 438–444 (2002)
18. Russell-Rose, T., Stevenson, M., Whitehead, M.: The reuters corpus volume 1-from yesterday's news to tomorrow's language resources (2002)
19. Sebastiani, F.: Machine learning in automated text categorization. ACM Comput. Surv. (CSUR) **34**(1), 1–47 (2002)

20. Sun, Z., Liu, X., Hu, K., Li, Z., Liu, J.: An efficient multi-label SVM classification algorithm by combining approximate extreme points method and divide-and-conquer strategy. IEEE Access **8**, 170967–170975 (2020)
21. Vidhya, A.: Multi label classification: solving multi label classification problems (2020). https://www.analyticsvidhya.com/blog/2017/08/introduction-to-multi-label-classification/. Accessed 23 Dec 2020
22. Wu, G., Zheng, R., Tian, Y., Liu, D.: Joint ranking SVM and binary relevance with robust low-rank learning for multi-label classification. Neural Netw. **122**, 24–39 (2020)
23. Yang, Y., Liu, X.: A re-examination of text categorization methods. In: Proceedings of the 22nd Annual International ACM SIGIR Conference on Research and Development in Information Retrieval, pp. 42–49 (1999)
24. Zhang, M.L., Li, Y.K., Liu, X.Y., Geng, X.: Binary relevance for multi-label learning: an overview. Front. Comp. Sci. **12**(2), 191–202 (2018)

A Scene Tibetan Text Detection by Combining Multi-scale and Dual-Channel Features

Cairang Dangzhi[1,2,3], Heming Huang[1,2,3(✉)], Yonghong Fan[1,2,3], and Yutao Fan[1,2,3]

[1] School of Computer, Qinghai Normal University, Xining 810008, China
2056617785@qq.com
[2] State Key Laboratory of Tibetan Intelligent Information Processing and Application, Xining 810008, China
[3] Key Laboratory of Tibetan Information Processing, Ministry of Education, Xining 810008, China

Abstract. Tibetan text detection in scenes plays a vital role in various applications, including image search, real-time translation, and the preservation of Tibetan cultural heritage. However, recognizing Tibetan text in natural scene images is a challenging task due to factors such as variable fonts, complex backgrounds, and poor imaging conditions. In this study, we present a novel approach called Multi-Scale Dual-Channel Feature Fusion (MDFF) for Tibetan scene text detection. Our method aims to accurately infer text in complex scenes by leveraging multi-scale interactions between texts. MDFF incorporates a feature pyramid network with skip connections, enabling the fusion of features at different scales in a hierarchical manner. Additionally, we employ a dual-channel attention (DCA) mechanism to capture rich interactions between text instances while mitigating the impact of background noise. Experimental results on the scene Tibetan text detection database (STTDD) demonstrate the effectiveness of MDFF, achieving an impressive F1 score of 85.20%. Our proposed method outperforms the baseline model by 5 percentage points and surpasses the performance of six state-of-the-art methods in single Tibetan text detection.

Keywords: Multi-scale Feature · Dual-channel Attention · Scene Tibetan Text Detection · Skip Connections · YOLO

1 Introduction

Text detection plays a crucial role in various applications, including text recognition in video, ancient documents, and real scenes. In the context of Tibetan text, methods for detecting text in clean printed images have shown promising results, owing to the maturity of Tibetan OCR systems. However, these methods face limitations when applied to real scenes due to challenges posed by complex backgrounds, diverse shapes, colors, fonts, and other factors. Additionally, text in real scenes can be occluded and blurred to varying degrees, further complicating the detection process. As a result, identifying Tibetan text in real scenes remains highly challenging.

H. Han and E. Baker (Eds.): SDSC 2023, CCIS 2113, pp. 158–171, 2024.
https://doi.org/10.1007/978-3-031-61816-1_11

In the field of traditional printed Tibetan text detection, text regions are typically identified by analyzing the connected regions of text pixels [1, 2]. However, this approach falls short when it comes to detecting Tibetan characters in natural scenes. The detection of Tibetan language text in natural scenes is still in its early stages of research. In recent years, deep learning-based detection methods have been increasingly utilized for detecting Tibetan ancient documents and text in natural scenes [3]. In a study by Song Hong et al. [4], a scene Tibetan detection method based on segmentation and component connectivity was employed. This method achieved a maximal F1 score of 72% in detecting Uchen fonts of Tibetan text in natural scenes. However, it exhibited shortcomings such as missed detections, false detections, and slow processing speed. Consequently, there is ample room for improving the performance and speed of Tibetan text detection methods in natural scenes.

In this paper, we propose a novel method called MDFF (Multi-Scale and Dual-Channel Feature Fusion) for Tibetan text detection in real scenes. Our method makes two key contributions. Firstly, we incorporate skip connections between feature layers, enabling the fusion of features from different layers with varying resolutions. This enhances the capability of feature extraction and improves the network's attention on text regions. Secondly, we introduce a dual-channel attention module at the front of the YOLOv7 detection head module. This module effectively prevents the loss of pixel information as the network depth increases. Experimental results on the STTDD (Scene Tibetan Text Detection) database demonstrate the effectiveness of MDFF, achieving an impressive F1 score. Moreover, our method exhibits a high Frames Per Second (FPS) score of 38 on the GeForce RTX 2060 GPU.

2 Related Works

In recent years, with the rapid development of artificial intelligence, many deep learning-related scene text detection methods have emerged [5]. These methods can be divided into three main categories: segmentation-based methods, component connection-based methods, and bounding box regression-based methods.

2.1 Segmentation-Based Text Detection Methods

The segmentation-based text detection methods generate compressed text segmentation mappings with different scales by detecting each text instance with the corresponding kernel at first, and then gradually expand the kernels to generate the final instance segmentation mapping.

U-Net, Mask R-CNN, and FCN are typical representatives [6]. To extract the backbone features more effectively, the feature pyramid networks [7] are employed to fuses high-level semantic information with low-level semantic information. Where, the high-level semantic information is rich in content but low in resolution, and it is difficult to store the object location information accurately; while the low-level semantic information contains less content, and the object location information is stored more accurately due to the high resolution.

The segmentation-based methods are good at detecting skewed text, curved text, or arbitrarily shaped text [8]. Especially, they are good at detecting the Tibetan-Chinese scene text [9]. However, they need a good many of complex post-processing steps that consume a lot of computational resources [10] and seriously affects the speed of text detection. Therefore, some researchers try to avoid using feature pyramid nets because of large computational efforts.

2.2 Component Connection Based Text Detection Methods

These kinds of methods start by segmenting an image into some connected components, regions of the image that are connected by pixels with similar properties. The connected components are then analyzed based on their spatial relationships with each other. Text regions are identified by looking for connected components with specific spatial arrangements, such as horizontal alignments, vertical alignments, or clusters of connected components with similar orientations.

Connectionist Text Proposal Network (CTPN) is the first of its kind that uses deep neural networks to predict and connect text fragments of a scene [11]. To detect arbitrarily shaped text, some components are used to locate characters at first, and then these components are connected.

The disadvantage of this method is that it cannot obtain rich relationships between text components, and it is not conducive to the division of text instances. To address these problems, in [12], Shixue Zhang et al. propose a new unified relational inference graph convolutional network. It can infer the possibility of links between text components and their neighboring components at first, and then, based on the inference results, all text components are aggregated into the whole text instance.

2.3 Bounding Box Regression Based Text Detection Methods

The text detection methods based on bounding box regression realize classification by regressing the parameter information of text boxes directly. These kinds of methods can be further divided into two types: single-stage detection methods and two-stage detection methods.

The single-stage detection methods estimate candidate targets directly, and they do not rely on region suggestions. Typical networks are the YOLO series and the single lens multi-box detectors [13].

The two-stage detection methods rely on region proposals, and the most representative method is Faster R-CNN [14]. Faster R-CNN is used to detect printed Tibetan text, Tibetan text in natural scenes, and the layout of modern books. The F1 score of detecting Tibetan text in natural scenes is 64.12% [15], and there is much more room for improvement.

3   Multi-scale and Dual-Channel Feature Fusion Method for Scene Tibetan Text Detection

The proposed method MDFF focuses on the multi-scale feature fusion and the improvement of attention of YOLOv7 [16] model. These improvements allow the model extracting features from local text regions more effectively, and thus improving the abilities of recognizing Tibetan scene text.

3.1   Framework of MDFF

Figure 1 shows us the framework of the proposed MDFF. It consists of four fundamental modules: the feature extraction module, skip connection multi-scale module, two-channel attention, and detecting head.

The layers $C_n(n = 1, 2, \ldots, 5)$ and $F_n(n = 2, 3, \ldots, 5)$ constitute the backbone network. Features are retrieved by the backbone network at first, and then improved by two-channel attention, and finally output by four detection heads as two categories of scene: Tibetan text (BO) and other language text (OL).

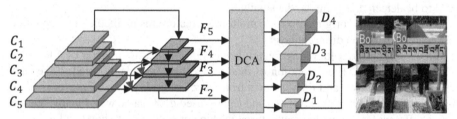

Fig. 1. Framework of MDFF. Where C_i represents the i-th downsampling layer, F_i represents the i-th upsampling layer, D_i represents the i-th detection head.

3.2   Feature Extraction

To each candidate box, MDFF extracts coordinate information, confidence scores, and class probabilities. Firstly, the input image is divided into $S \times S$ windows of the same size, and, for each window, B candidate boxes are generated. The predicted parameters for each candidate box include four coordinates (x, y, w, h) and confidence score information c. Secondly, the class probabilities for each candidate box that contains text objects are calculated. And finally, each box generates a vector with a dimension of B × 5 + N, where B represents the number of candidate boxes, N represents the number of categories, and the number 5 represents the 4 coordinates and the confidence information c. As shown in Fig. 2, a $S \times S \times (B \times 5 + N)$ tensor is generated.

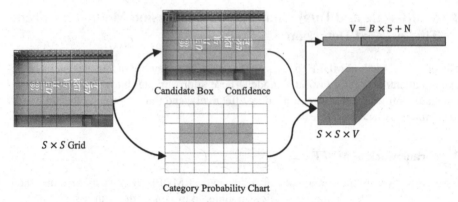

Fig. 2. The procedure of Feature Extraction

3.3 Skip Connections of Multi-scale Feature Fusion

When the number of layers of convolutional neural networks increase, it loses the pixel features. The missed pixels have important impacts on the location of the objects that need to be detected, especially, the small targets.

To preserve more original pixels, a multi-scale feature fusion method is proposed to fuse features between different layers.

Firstly, the low-resolution F_5 feature layer is upsampled by a factor of 4 to obtain a new feature layer. The resolution of this new feature layer is adjusted to the same resolution as F_3 layer, and then features from new feature layer and F_3 layer is fused. Secondly, features from the layers F_4 and F_2 are fused in the same way. Finally, features across different layers are connected through skip connections, as shown in Fig. 3.

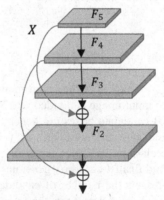

Fig. 3. Skip Connect of Multi-scale Feature Module. X denotes feature; F_5, ..., F_2 denotes the layers in upsampling in the backbone network.

The main idea of the skip connect is to add the input of a layer to its output, so that the network can learn residual information instead of directly learning the original features. Let the input be denoted as X and the output of layer l be denoted as y_l. Then, the formula for the residual connection is:

$$y_l = F(X_l) + X_l. \tag{1}$$

where, $F(X_l)$ represents the feature extraction process of layer l and X_l represents the input of layer l. The addition operation is performed elementwise, meaning the elements at corresponding positions are added.

3.4 Dual-Channel Attention

The attention is calculated on two feature domains, namely, channel domain [17] and spatial domain [18]. It can effectively suppress background noise and enhance the features of foreground text areas, and thus enable the detection head to extract more useful features. The proposed dual-channel attention is shown in Fig. 4.

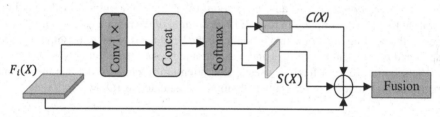

Fig. 4. Dual-channel Attention. F_i denotes feature of the i-th layer upsampling output.

Feature fusion refers to combining multiple features into one feature to improve the efficiency and accuracy of the model. Among them, add and concat are two common feature fusion methods.

Firstly, to enhance the non-linear characteristics, a 1×1 convolution operation is applied to the F_i layer to generate a new feature layer. The purpose is to maintain the same number of channels as the previous layer while keeping the feature map resolution unchanged, so that the network can learn more complex feature information.

Secondly, the new feature layer and the F_i layer are concatenated using the concatenation (concat) method, and the calculation formula is provided in Eq. (2).

$$X = [X_{in1}, X_{in2}], \tag{2}$$

where X_{in1} and X_{in2} are the two input features and X is the output feature. In concat, the two input features are concatenated along a certain dimension to form a higher-dimensional feature. For example, if X_{in1} has a shape of $(batch\_size, n1)$ and X_{in2} has a shape of $(batch\_size, n2)$, then the shape of X after concatenation is $(batch\_size, n1 + n2)$. The number of channels in the feature X after the concatenation is equal to $n1 + n2$, which is twice the number of channels in the original feature layer.

Subsequently, the Sigmoid function is used to activate the features of X. The features of X is divided into two groups to create two feature maps. Channel attention is used to suppress background interference for the first set of features. Then the Channel attention is calculated as follow:

$$C(X) = \frac{1}{1 + e^{-(w_1(w_0(X_{avg}^c)) + w_1(w_0(X_{max}^c)))}},\tag{3}$$

where X is the input feature map, c is the number of channels, w_0 and w_1 are the weights of multi-layer perceptron, which are shared for both inputs. avg And max are operations for average pooling and max pooling, respectively. Pooling operation is a commonly used downsampling operation in convolutional neural networks. The formula for average pooling and max pooling can be written as:

$$average\,pooling(X)_{i,j,k} = \frac{1}{w \times h}\sum\nolimits_{a=0}^{h-1}\sum\nolimits_{b=0}^{w-1}X_{i,j \times s+a,k \times s+b},\tag{4}$$

$$max\,pooling(X)_{i,j,k} = \underset{a,b}{max}X_{i,j \times s+a,k \times s+b},\tag{5}$$

where $average\,pooling(X)_{i,j,k}$ and $max\,pooling(X)_{i,j,k}$ are the outputs of the pooling operation for the i-th sample, j-th row, and k-th column of the input tensor X, and s is the stride of the pooling operation, and w and h are the height and width of the input feature map, a and b represent the vertical and horizontal indices of the pooling window.

The second group of features uses spatial attention to enhance the features of the foreground text area. Then the spatial attention is calculated as follow:

$$S(X) = \frac{1}{1 + e^{-(f^{7 \times 7}([X_{avg}^s, X_{max}^s]))}},\tag{6}$$

where $f^{7 \times 7}$ is convolutional window size, and s is the Spatial feature dimension.

Ultimately, the features of channel attention, spatial attention, and original feature map layer output are fused using the add method. The calculation method is as follows:

$$Z_a = \sum\nolimits_{i=1}^{c}C_i + \sum\nolimits_{i=1}^{c}S_i + \sum\nolimits_{i=1}^{c}F_i,\tag{7}$$

where C is the feature of the channel attention layer output, S is the feature of the spatial attention layer output, and F denotes feature of the i-th layer upsampling output.

3.5   Alpha-IoU Loss Function

The effectiveness of text detection depends greatly on the definition of the loss function which measures the accuracy of the model in predicting the target location results. The text position on the image can be represented by four-dimensional vectors (x, y, w, h), whose elements represent the center point coordinates and width height of the text bounding box. In the process of modeling text location, the model needs to find a mapping relationship that uses this relationship to make the predicted text boxes infinitely close to the real text boxes, which is as follows:

$$f\left(P_x, P_y, P_w, P_h\right) = (\tilde{G}_x, \tilde{G}_y, \tilde{G}_w, \tilde{G}_h) \approx (G_x, G_y, G_w, G_h).\tag{8}$$

where P represents predicted text boxes, $\tilde{G}$ represents generated candidate boxes, G represents real text boxes.

The bounding box regression model utilizes translation and scaling transformations to achieve mapping. Among them, the translation transformation is:

$$\begin{cases} \tilde{G}_x = P_w d_x(P) + P_x \\ \tilde{G}_y = P_h d_Y(P) + P_y \end{cases} \tag{9}$$

and the scale transformation is:

$$\begin{cases} \tilde{G}_w = P_w exp(d_w(P)) \\ \tilde{G}_h = P_h exp(d_w(P)) \end{cases} \tag{10}$$

The traditional calculation method for *IoU loss* is:

$$IoU\ loss = -\ln\frac{|G \cap P|}{|G \cup P|}, \tag{11}$$

if the prediction box P and real text box G do not intersect, the loss is 0, and the model cannot learn. To optimize the model, the alpha-IoU boundary regression loss function is employed [19], and it is calculated as in Eq. (12):

$$L_{\alpha-IoU} = \frac{1 - IoU^{\alpha}}{\alpha}, \alpha > 0. \tag{12}$$

The trade-off parameter alpha in this function controls the weight comparison between the intersection-to-union ratio and Focal Loss, typically ranging from 0.25 to 0.75. A value less than 0.5 may be better when the target category is imbalanced, while a value of 0.5 or greater can be considered when the target category is relatively balanced. The specific values need to be adjusted according to the actual situation. The value used in this article is 0.5.

4 Experiments

4.1 Tibetan Scene Text Dataset

Currently, there is no publicly available natural scene Tibetan text detection database. Therefore, a Scene Tibetan Text Detection Database (STTDD) is constructed.

The STTDD database contains a total of 500 images, all with a size of 640 × 640. There are 2159 text target entities in the database. All target entities are divided into small, medium, and large groups: 369 small-sized text targets, 739 medium-sized text targets, and 1051 large-sized text targets. The ratio of three groups is roughly 1:2:3, which meets the requirements of creating a standard target detection database.

The dataset divided into a training set of 400 samples and a test set of 100 samples in an 8:2 ratio. There are 1642 text target entities in the test set and 517 text target entities in the test set. As shown in Fig. 5, the samples of STTDD database.

(a) Horizontal text   (b) Ambiguous text

(c) Slanted text   (d) Bending text

Fig. 5. Some samples of STTDD

4.2 Results Analysis

To assess the efficacy of MDFF, an ablation analysis of SCMF and DCA is conducted at first, and then the performance of MDFF is compared with those of previous research.

4.2.1 Ablation Studies

To demonstrate the effectiveness of SCMF and DCA, Table 1, below, lists and compares the F1 scores of models YOLOv7, chosen as the baseline, YOLOv7 + SCMF, YOLOv7 + DCA, YOLOv7 + SCMF + DCA. All have the same size of hyperparameters.

Table 1. Ablation analysis of SCMF and DCA

Models	F1(%)
YOLOv7 (baseline model)	67
YOLOv7 + SCMF	69
YOLOv7 + DCA	71
YOLOv7 + SCMF + DCA(Ours)	72

Note: In the YOLOv7 model, the calculation result of F1 scores is, by default, integers. (See Fig. 6)

To the test set of STTDD, the F1 score of baseline model achieves 67%. It increases to 69% when the module SCMF is used; it increases to 71% when the module DCA is used;

and it increases further to 72% when both are used. The proposed method outperforms the baseline model by 5 percentage points.

To better illustrate the effectiveness of the above-mentioned 4 models on text detection, their F1 scores are shown in Fig. 6.

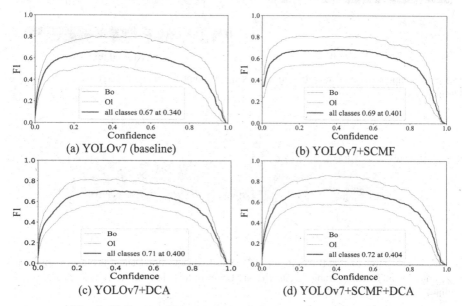

Fig. 6. F1 Scores of Four Methods of Tibetan Scene Text Detection

Figure 6 tell us that the SCMF module is more effective in detecting text regions of other language (See Fig. 6 (b)). Since these regions contain more small targets, and the multi-scale feature fusion method is good at detecting small targets. The DCA method, which uses attention-enhanced feature extraction, is effective for detecting both Tibetan texts and texts in other languages (See Fig. 6(c)). Therefore, the proposed method, which uses both SCMF and DCA, achieves the best results on the test set (See Fig. 6(d)).

As shown in Fig. 6, the F1 scores of the four models for detecting Other Languages (OL) are all below 60%. The main reason is that the images in the Other Languages category contain many small text targets (See Fig. 7 (a)), and the detection of which is more challenging.

In the test sample, in addition to the text of target language, some small sticker advertisements also contain many texts of other language, as shown in Fig. 7 (a). The red arrow points to the enlarged version. The text on these small stickers is very small and they are almost undetectable by the human eye. By comparing, Fig. 7(b) show us the text in normal size. To the proposed detection method, some of small targets would be missed.

Figure 8 shows us the sizes distribution of all text boxes in the STTDD database. The number of small target text boxes in in other languages is great than those in Tibetan, and it makes small target text difficult to be detected in other languages.

(a) Small target text   (b) Normal text

Fig. 7. Examples of small target. Red arrows in (a) point to enlarged versions. (Color Figure Online)

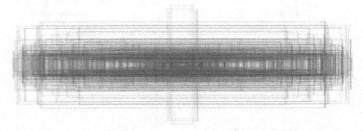

Fig. 8. Visualization of Text Boxes in Scene Tibetan Text Detection Data. The blue rectangles are boxes of texts in Tibetan; Orange rectangles are boxes of texts in other languages. (Color Figure Online)

4.2.2   Comparison of Performance with Other Research Results

The comparison analysis in this section focusses mainly on the different methods and whether the pre-training is used. The performance of these methods is evaluated by such metrics as precision, recall, and the F1 score. The results are shown in Table 2, where OS denotes single-level detection methods, TS denotes two-level detection techniques, Seg denotes segmentation-based detection methods, CC denotes connected component text detection, and Box denotes bounding box regression-based detection methods.

The experimental results of the literature [9] in Table 2 are special because no single Tibetan text scene detection results is provided. Instead, an F1 score of 73.88% is obtained from a testing set with 200 images. These images are selected from the processing where a pre-training with the IC17-MLT (9000) multi-lingual database is conducted at first and then a fine-tuning with 325 real-scene Sino-Tibetan bilingual data is done. The F1 scores of other techniques are results of single Tibetan text scene detection. The proposed MDFF obtains an F1 score of 85.2%. It brings a 13.20 percentage points increase over the EAST framework proposed by Hong Song et al.

Table 2. The comparison between MDFF and other scenes Tibetan text detection methods

Literature	Methods	Pre-training	P(%)	R(%)	F1(%)
Song Hong, et al. [4]	DBNet (Seg)	No	89.00	59.00	71.00
Song Hong, et al. [4]	CTPN (CC)	No	45.00	34.00	39.00
Song Hong, et al. [4]	EAST (Seg)	No	83.00	64.00	72.00
Jincheng Li [9]	Seg	Yes	75.47	72.36	73.88
Yanru Wu [15]	Box (TS)	No	52.33	82.79	64.12
Song Hong [20]	DBNet (Seg)	No	87.00	55.00	67.00
MDFF model (Ours)	Box (OS)	No	92.20	79.20	85.20↑13.20

Note: The first three models are the results of experiments in the same literature

5 Discussion

To improve the detection effect of Tibetan and foreign languages in natural scenes, a multi-scale and dual-channel feature fusion-based method, called MDFF, is proposed. There are still some challenges in the text detection task such as small targets detection and arbitrarily shaped text detection.

5.1 Small Target Text Detection

In target detection tasks, whether in text detection or other entity target detection, small target detection is a difficult task due to low resolution and few features. Therefore, in future work, more feature layer fusion will be used, and some weight coefficient will be introduced to change the response value of small targets when the loss is calculated.

5.2 Arbitrary Shape Text Detection

Arbitrary shape text detection is another difficult problem in scene text detection. Usually, a bottom-up approach is used to solve this problem and it can fit the boundaries of curved text and skewed text better. However, complex post-processing often reduces detection speed.

Although the proposed method can detect arbitrarily shaped text, however, there are a lot of redundant background pixels, as shown in Fig. 5(d). In future work, we will improve the model by considering the tilt angle of the text and redefining the loss function to improve the ability to detect arbitrarily shaped text.

6 Conclusion

In this study, we propose a novel one-stage scene Tibetan text detection (MDFF) method that achieves improved accuracy in detecting texts in both Tibetan and other languages. To enhance the text detection capabilities, we introduce two key modules: the Scene

Context-Maintaining Fusion (SCMF) module and the Discriminative Context Attention (DCA) module. The SCMF module plays a crucial role in preserving original pixel features, thereby enhancing the ability to detect small target text. By maintaining the contextual information within the scene, this module significantly contributes to the overall performance of the detection system. Additionally, the DCA module effectively suppresses interference from complex backgrounds, allowing for enhanced feature extraction from foreground text regions. By focusing on discriminative context attention, this module helps in isolating and emphasizing the essential characteristics of the text, leading to improved detection accuracy. The MDFF method achieves an impressive F1 score of 85.2% on the test set of STTDD. Moreover, when running on the GeForce RTX 2060 GPU, it achieves a favorable Frames Per Second (FPS) score of 38, indicating its computational efficiency. In the future, our research will continue to explore and delve deeper into text detection for small target text and text of arbitrary shapes. These areas present unique challenges that warrant further investigation, and we are committed to advancing the field by developing innovative solutions to address them.

Acknowledgement. The authors acknowledge National Natural Science Foundation of China (Grant: 62066039), Natural Science Foundation of Qinghai Province (Grant: 2022-ZJ-925), and the "111" Project (D20035).

References

1. Yuehui, H.: Design and Implementation of Tibetan Ancient Book Recognition System. Northwest Minzu University (2019)
2. Yang, C.: Design and Implementation of Printed Tibetan Recognition Software on Android Platform. Northwest Minzu University (2020)
3. GongQuZhuoMa, C.S.: Text region detection of tibetan ancient books based on semantic segmentation[J/OL]. Comput. Simul. **39**(5), 448–454 (2022)
4. Song, H., Dingguo, G., SanPai, C., et al.: Detection and recognition of wujin style tibetan scripts in natural scenes. Comput. Syst. Appl. **30**(12), 332–3389(2021)
5. Mingxin, H., Yuliang, L., Zhenghao, P., et al.: SwinTextSpotter: Scene Text Spotting via Better Synergy between Text Detection and Text Recognition [OL]. http://arxiv.org/abs/2203.10209. Accessed on 20 Sep 2022
6. Jonathan, L., Evan, S., Trevor, D.: Fully convolutional networks for semantic segmentation. In: Proceedings of 2015 IEEE Conference on Computer Vision and Pattern Recognition (CVPR). Boston, MA, USA, pp. 3431–3440 (2015)
7. Tsung, Y.L., Piotr, D., Ross, G., et al.: Feature pyramid networks for object detection. In: Proceedings of 2017 IEEE Conference on Computer Vision and Pattern Recognition (CVPR), Hawaii, America, pp. 936–944. IEEE (2017)
8. Yuliang, L., Hao, C., Chunhua, S., et al.: ABCNet: Real-time Scene Text Spotting with Adaptive Bezier-Curve Network [OL]. http://arxiv.org/abs/2002.10200. Accessed on 20 Sep 2022
9. Jincheng, L.: Tibetan-Chinese Bilingual Natural Scene Text Detection and Recognition System. Northwest Minzu University (2021)
10. Fangfang, W., Yifeng, C., Fei, W., et al.: TextRay: Contour-based Geometric Modeling for Arbitrary-shaped Scene Text Detection [OL]. http://arxiv.org/abs/2008.04851. Accessed on 20 Sep 2022

11. Zhi, T., Weilin, H., Tong, H., et al.: Detecting text in natural image with connectionist text proposal network. In: Proceedings of European Conference on Computer Vision. Amsterdam, Netherlands: Springer, pp. 56–72. Springer, Heidelberg (2016)

12. Shixue, Z., Xiaobin, Z., JieBo, H., et al.: Deep relational reasoning graph network for arbitrary shape text detection. In: Proceedings of the 2020 IEEE/CVF Conference on Computer Vision and Pattern Recognition (CVPR). Seattle, USA, pp. 9696–9705. IEEE Computer Society Press (2020)

13. Liu, W., Anguelov, D., Erhan, D., et al.: SSD: single shot multibox detector. In: Proceedings of European Conference on Computer Vision. Amsterdam. Netherlands, pp. 21–37. Springer, Heidelberg (2016)

14. Ren, S., He, K., Girshick, R., et al.: Faster R-CNN: towards real-time object detection with region proposal networks. IEEE Trans. Pattern Anal. Mach. Intell. **39**(6), 1137–1149 (2017)

15. Yanru, W.: Research on layout detection technology of Tibetan modern printed. Tibet University (2020)

16. Chien, Y.W., Alexey, B., Hong, Y.M.L., et al.: YOLOv7: trainable bag-of-freebies sets new state-of-the-art for real-time object detectors [OL]. https://doi.org/10.48550/arXiv.2207.02696. Accessed on 06 June 2022

17. Jie, H., Shen, L., Albanie, S., et al.: Squeeze-and-excitation networks. IEEE Trans. Pattern Anal. Mach. Intell. **42**(8), 2011–2023 (2020)

18. Sanghyun, W., Jongchan, P., Joon-Young, L., et al.: CBAM: convolutional block attention module. In: Proceedings of the European Conference on Computer Vision. Munich, Germany, pp. 3–19. Springer, Heidelberg (2018)

19. Jiabo, H., Sarah, E., Xingjun, M., et al.: Alpha-IoU: A Family of Power Intersection over Union Losses for Bounding Box Regression [OL]. https://arxiv.org/abs/2110.13675. Accessed on 22 Jan 2022

20. Hong, S.: Research on the Detection and Recognition Method of Wujin Tibetan Script in Natural Scenes. Tibet University (2021)

DGL Version 2 – Random Testing in the Mobile Computing Era

Peter M. Maurer[✉]

Department of Computer Science, Baylor University, Waco, TX 76798, USA
`Peter_Maurer@baylor.edu`

Abstract. DGL has been a popular tool for testing software and hardware. The original was file based, meaning that it creates files of test data. This method of data generation does not work for embedded applications and applications for hand-held systems. The new version of DGL, Version 2, has been enhanced to be a better tool for modern applications. The new version has improved deployment features including direct embedding in C++ code. It also has improved input and output facilities, which are described here.

1 Introduction

DGL, the Data Generation Language [1, 2], has proven to be a useful tool for testing both software and hardware. As far as we know, the approach used by DGL is unique. There are a number of specialized data generation tools, such as [3–7], but as far as we've been able to determine, DGL is the only general-purpose data generation tool. DGL is based on the concept of probabilistic context free grammars [8]. The data to be generated is described using an extended context free grammar. A data generator program is created automatically from the grammar and is subsequently used to generate complex randomized data. The original version of DGL (DGL Version 1) was designed to be a self-contained system with no dependencies on outside sources, such as files or data bases. Figure 1 shows the basic structure of DGL Version 1.

DGLVersion 1

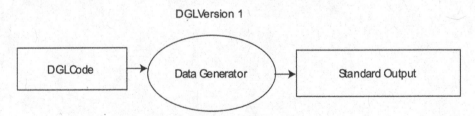

Fig. 1. Input and Output Data, Version 1.

The DGL code is compiled into the high-level language C, which is then compiled into the data generator program. All generated data items must appear somewhere in the

H. Han and E. Baker (Eds.): SDSC 2023, CCIS 2113, pp. 172–183, 2024.
https://doi.org/10.1007/978-3-031-61816-1_12

DGL code. When DGL was first created, most computer programs were file processors. A program would read one or more input files and create one or more files as output.

Things have changed considerably since DGL was first created. File processors still exist, but most programs (or apps, as they are now called) are either embedded or interactive. Mainframes and desktop computers are no longer the mainstay of computer programming. Today, tablets, cell phones, and other smart devices are the mainstay. Although DGL still has enormous potential for testing, simulation, .and other problems requiring random data, the paradigm of creating files of input data is no longer useful.

DGL Version 2 was developed to be a tool for the modern programming world. Self-contained data is no longer realistic. Generated data must come from several sources, including data bases and external files. The current version of DGL (Version 2) has several mechanisms for data input and for data output. Other mechanisms are in place to include the data generator as part of the application. The data generation software will always be present and will be activated using conditional compilation or other mechanisms.

2   DGL Structure

The basic unit of the DGL language is the production. An example of a production is given in Fig. 2. This is a *normal* production, corresponding to a set of productions in an ordinary context free grammar. A production starts with a unique name, followed by key words and a list of alternatives. The normal production has no key words.

Ordinary Context free grammar:
$\langle Color \rangle \rightarrow red$
$\langle Color \rangle \rightarrow blue$
$\langle Color \rangle \rightarrow green$
$\langle Color \rangle \rightarrow orange$
$\langle Color \rangle \rightarrow yellow$
$\langle Color \rangle \rightarrow pink$

Equivalent DGL:
Color: red, blue, green, orange, yellow, pink;

Fig. 2. CFG to DGL.

In DGL non-terminals are specified with a leading percent sign: %{a}, or %{Color}. (If the production name is a single letter, the braces may be omitted, as in %a.) The default start symbol of a DGL grammar is "main". In DGL Version 1, a file of data is generated by interpreting the non-terminal %{main} 100 times. Each interpretation of %{main} produces a single item of data.

A string is interpreted by scanning it from left to right, and replacing each occurrence of a non-terminal with a selection from the alternatives of the corresponding production.

When a selection is made from a normal production, one of the alternatives of the production is chosen with equal probability and with replacement. Thus, for the "Color" production of Fig. 2, each of the six colors is chosen with probability 1/6, and a particular color may be chosen several times.

Keywords are used to alter the behavior of the production. The two productions of Fig. 3 are examples. For the first production Color1 the alternatives are chosen without replacement. For the second production Color2 the alternatives are chosen sequentially. Additional keywords are available to determine what happens when the production runs out of alternatives. When we speak of production types we are most often referring to the keyword that defines them. Thus Color1 from Fig. 3 is a "unique" production and Color2 is a "sequence" production

> Color1: **unique** red, blue, green, orange, yellow, pink;
> Color2: **sequence** red, blue, green, orange, yellow, pink;
> V: **variable**;

Fig. 3. Altered Selection Methodology.

The production "V" of Fig. 3 is an example of the most powerful production, the "variable" production. The variable "V" acts like a normal production with a single alternative. However, the single alternative can be dynamically changed during the interpretation of other productions. Figure 4 gives an example of assigning a value to a variable.

> V: variable;
> main: %{Color.V}%{V}%{V}%{V};
> Color: red, blue, green, orange, yellow, pink;

Fig. 4. Variables.

The non-terminal %{Color.V} is known as an "active" non-terminal. Active non-terminals are used to give access to additional features of productions. For variables such as "V" active non-terminals are used to assign values to variables. For %{Color.V}, a selection is made from the alternatives of "Color", the selection is completely interpreted, and the result is assigned to "V". In the example of Fig. 4, suppose that "red" is assigned to V by the non-terminal %{Color.V}. The remaining three non-terminals will produce the string "red", and the production "main" will produce the string "redredred". The addition of variables to DGL makes DGL universal in the theoretical sense.

3 Embedded Code

Dgl specifications may be directly incorporated into C++ code by using the tags < dgl > and < /dgl >. These tags must be on a separate line and must begin at the extreme left. Anything between these two tags is treated as dgl specifications. The source file must be filtered through the dgl compiler before being compiled by the C++ compiler.

The main function of embedded dgl is to assign random values to variables and objects. Any cvar production specified in the embedded code generates a random assignment to the specified variable (See Sect. 5.2). Non-cvar productions are resource productions used by the cvar productions. Embedded code can be used to initialize variables, or it can be placed inside a loop to generate successive random values.

The following is an example of embedded dgl.

```
int main()
{
  char *stuff;
  for (int i=0 ; i<10 ; i++)
  {
<dgl>
    stuff: cvar "abc","def","ghi","jkl","mno","pqr","stu";
</dgl>
    cout<<stuff;
    free(stuff);
  }
}
```

4 Database

DGL database access is described in [9]. Four productions have been added to DGL to control database access. The database production connects to a database, the dbtable and dbquery productions provide access to database tables and SQL queries. The dbfield productions provide access to the fields of the current record. Databases may be used for both input and output.

5 Input and Output

In addition to being compiled into stand-alone data generation programs, Version 2 DGL code can be compiled into data-generation subroutine. Data generation subroutines are linked with an application. Both embedded DGL and data generation subroutines permit input and output through global variables. For embedded DGL, local variables can also be used for input or output as long as the embedded DGL is within the scope of the variables.

Figure 5 shows the flow of data into and out of a Version 2 data generator. As with Version 1, the data generator is created automatically from a DGL grammar. In Version 1, data is generated on the standard output, which is typically redirected to a file.

5.1 Input from C + + Variables

Version 2 preserves the mechanisms of Version 1. Thus, one source of data is the DGL code itself. This is demonstrated in the example of Figs. 2 and 3, where color data is

DGLVersion 2

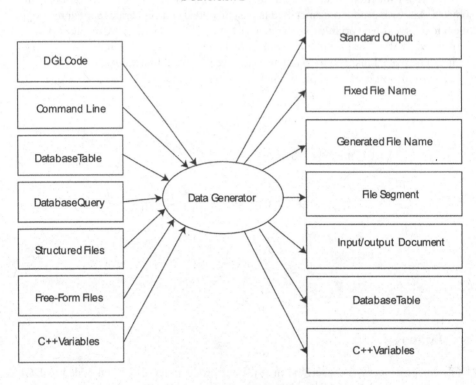

Fig. 5. Input and Output Data Version 2.

taken from the alternatives of the "Color" production. In most cases, at least some of the generated data comes from production alternatives.

In Version 2, a major source of input data is C+ + global variables. The "icvar" production provides a hook between DGL productions and C+ + variables. There are several different versions of this production to handle different types of variables.

Regardless of the data type, each icvar production begins with a name, a colon, and the "icvar" keyword. Normally the name of the C+ + variable is identical to the name of the production, but if this is not desirable, the C+ + variable name is enclosed in parentheses following the icvar keyword. Figure 6 gives an example.

V1 : **icvar** ;
V2 : **icvar(ClientName)** ;

Fig. 6. C + + input variables.

For the productions of Fig. 6, the non-terminal %{V1} produces the contents of the global C + + variable "V1". The non-terminal %{V2} produces the contents of the CD + + variable "ClientName". Since no type has been specified, the default type of "char *" is assumed. The contents of V1 and ClientName are assumed be null terminated character strings. These strings are passed through the interpreter, and any non-terminal symbols are replaced with selections from the appropriate productions.

Character data can also be input using the "char" keyword, as shown in Fig. 7.

X1: **icvar char (25)** ;
X2: **icvar(Description) char (25)** ;

Fig. 7. Fixed-length strings.

The char keyword is used to specify fixed length strings. For the productions of Fig. 7, the non-terminal %{X1} produces the first 25 characters of the global variable X1, while the non-terminal %{X2} produces the first 25 characters of the global variable Description.

Numeric data can be input from binary integers and floating point numbers. The keywords binary, ubinary, float, double, and longdouble are used for this purpose. Figure 8 gives examples.

Binary Integers.

G1: **icvar binary(8);**
G2: **icvar ubinary(32);**
G3: **icvar float;**
G4: **icvar double;**
G5: **icvar longdouble;**

The specified variable is assumed to be a char, short, int, or long long, depending on the < length > specification. 1 through 8 specify a char, 9 through 16 specify a short, 17 through 32 specify an int, and 33 through 64 specify a long long. Anything larger than 64 is assumed to be 64. If the length is omitted, it is assumed to be 32. If the **ubinary** keyword is used, the variable is assumed to be unsigned, if the **binary** keyword is used, it is assumed to be signed. When a non-terminal such as %{G1} is interpreted, the binary integer is converted to decimal ASCII, and then processed as a string.

For the keywords float, double, and longdouble, the specified variable is assumed to be a float, double or long double variable respectively. The value of the variable is converted to ASCII, with up to 15 decimal places, and is then processed as a string.

Boolean input can be obtained using the bool and cbool kewords. The specified variable is normally assumed to be a bool, but it also may be an integer variable of any size. If the keyword **bool** is used, the value of the variable is converted to the string "1" or the string "0", depending on whether the variable is true or false respectively. If the variable is an integer, any non-zero value is converted to the string "1", while zeros are converted to the string "0". If the keyword **cbool** is used, the values "1" and "0" are converted to is converted to the strings "true" and "false" respectively. Figure 8 gives examples.

> B1 :icvar bool ;
> B2 :icvar cbool ;

Fig. 8. Boolean Input.

5.2   Output to Global Variables

The cvar keyword is used to link a production to a global variable. Figure 9 shows some examples of cvar productions. The name of the global variable matches the production name, unless a different name is specified after the cvar keyword. The productions X1 and X2 output null terminated strings to the variables X1 and Myspace::MyName respectively. Production X3 outputs a binary floating point number to the variable X3.

> X1: cvar abc,def,ghi;
> X2: cvar("MySpace::MyName") Fred, Joanne, Mary;
> X3: cvar float 3.14159,2.5,4;

Fig. 9. Cvar examples.

Figure 9 shows output to simple variables. The output is performed by interpreting the non-terminals %{X1}, %{X2}, and %{X3}. Output can also be to an array. Every element of the array is filled by a single interpretation of the production's non-terminal. Arrays are specified using the "dim" keyword. When the dim keyword is used, the array may have any number of dimensions. There are three different forms of the dim specification, as Fig. 10 demonstrates.

> Type 1:
> **dim**(3)
> **dim**(3,4,5)
> **dim**(6,7)
>
> Type 2:
> **dynamic dim**(7,5),
> **dynamic dim**(8)
>
> Type 3:
> **dynamic dim**(3,4,2) **dimtab** (DimList)

Fig. 10. The dim specification.

If the first form of the specification is used, the array is assumed to be statically allocated. The proper number of dimensions must be supplied in the declaration, and the dimensions specified in the dim specification must match the dimensions specified in the variable. Figure 11 shows how one, two, and three dimensional arrays must be declared.

```
int One[3];           // dim(3)
int Two[3][4];                // dim(3,4)
int Three[3][4][5];   // dim(3,4,5)
```

Fig. 11. Array Dimensions.

If the **dynamic** keyword is present, the array will be created dynamically. The dimension numbers may contain non-terminals, causing the size of each dimension to be determined dynamically. Dynamically created arrays use a pointer-to-pointer type structure, so the variable declarations must include the proper number of indirect indicators, as in Fig. 12, which declare a one dimensional, a two dimensional, and a three dimensional array respectively.

```
int *one;
int **two;
int ***three;
```

Fig. 12. Dynamic Arrays.

When size of dimensions are determined dynamically, as in **dynamic dim**(%a,%b,%c) it is advisable to also use the **dimtab** specification, as in Fig. 13.

dynamic dim(%a,%b,%c) dimtab("DTab")

Fig. 13. The dimtab specification.

In Fig. 13, *DTab* must be the name of an array of int's whose dimension matches the number of dimensions in the **dim** specification. Thus *DTab* must be declared as follows. The elements of DTab will contain the size of each dimension after the array is filled with data.

int DTab[3];

If a cvar production is used to generate several selections for the same dynamic array variable, it is the responsibility of the user to free the previous value of the variable. Suppose a three dimensional array of integers has been created in the variable *a*, which has been declared as follows.

int ***a;

Suppose further that the dimensions are static: dim(4,5,6). Before a new value is generated for *a*, it is necessary to free the previous value using the following code.

```
for (int i=0 ; i<4 ; i++)
{
    for (int j=0 ; j<5 ; j++)
    {
        free(a[i][j]);
    }
    free(a[i]);
}
free(a);
```

When dimensions are dynamic, the dimension table should be used instead of a hard limit. Suppose the following specification is used: **dynamic dim**(%a,%b,%c) **dimtab**("DTab"). Then the following code must be used to free the previous value.

```
for (int i=0 ; i<DTab[0] ; i++)
{
    for (int j=0 ; j<DTab[1] ; j++)
    {
        free(a[i][j]);
    }
    free(a[i]);
}
free(a);
```

5.3 Output Data Types

Figure 14 gives examples of each of the available output types.

5.4 Random Variates in C + + Variables

Cvar productions are available for generating random variates for a number of different widely-used probability distributions. This data is all numeric. The term "random variates" implies that the generated numbers will have a particular probability distribution if a sufficient number of them are generated. Each of these productions permits a **dim** specification to be used, allowing a large number of random variates to be generated in a single operation. Figure 15 shows the various different probability distributions that are available. Unlike the previous cvar productions, no list of alternatives is present. A full discussion of random variate generation is given in [10].

Z1 **cvar** abc,def,ghi;
Z2: **cvar** char (5)<dim> abc,def,ghi ;
Z3: **cvar binary**(8) 1,2,3,4,5;
Z4: **cvar binary**(16) 1,2,3,4,5;
Z5: **cvar binary**(32) 1,2,3,4,5;
Z6: **cvar binary**(64) 1,2,3,4,5;
Z7: **cvar ubinary**(8) 1,2,3,4,5;
Z8: **cvar ubinary**(16) 1,2,3,4,5;
Z9: **cvar ubinary**(32) 1,2,3,4,5;
Z10: **cvar ubinary**(64) 1,2,3,4,5;
Z11: **cvar float** 3.1,2.2,6.125;
Z12: **cvar double** 3.1,2.2,6.125;
Z13: **cvar longdouble** 3.1,2.2,6.125;
Z14: **cvar bool** 0,1,2,true,false;

Fig. 14. Output Variable Types.

<name> :**cvar poisson** (%{p});
<name> :**cvar geometric** (%{p});
<name> :**cvar binomial** (%{p});
<name> :**cvar binomial** (%{p1},%{p2});
<name> :**cvar hypergeometric** (%{p1},%{p2},%{p3});
<name> : **cvar powerlaw** (%{p1},%{p2});
<name> : **cvar betavariate** (%{p1},%{p2});
<name> : **cvar negativebinomial** (%{p1},%{p2});
<name> : **cvar gammavariate** (%{p});
<name> : **cvar expvariate**;
<name> : **cvar expvariate** (%{p});
<name> : **cvar expvariate** (%{p1},%{p2});
<name> : **cvar norvariate**;
<name> : **cvar norvariate** (%{p});
<name> : **cvar norvariate** (%{p1},%{p2}) ;

Fig. 15. Random Variate Cvar Productions.

6 Files

File output has been expanded to include multiple output files, in addition to the standard output. The main mechanism for doing output to multiple files is the file production shown in Fig. 16.

Active non-terminals are used to control files. Assuming we have the productions of Fig. 16, the active non-terminal %{a.F1} makes a selection from production "a", completely interprets the selection, and writes the result to "abc.txt". F1 will respond to a number of other active non-terminals, such as %{$open.F1}, %{$close.F1} and

F1: **file** "abc.txt";
F2: **file** "mydir/observations.txt";
F3: **file** %{GenFileName};

Fig. 16. The File Production.

a few others. Under normal conditions, the file "abc.txt" will be opened and closed automatically without any action required by the user.

If the file F3 is closed and reopened, the %{GenFileName} production will be reinterpreted, allowing F3 to be used with several different files.

Any output file can be broken into separately generated sections. This feature is motivated by the idea of generating an executable program with global variables. The global declarations must appear in one section of the output and the code must appear in another section. Variables and code can be generated simultaneously using sectioned output. The first step is to declare the sections, as shown in Fig. 17. Output is directed into a section, S1, for example, using the non-terminal %{XMLOutput.S1}. A selection is made from the production XMLOutput. The selection is completely interpreted, and the result is stored in section S1. Each section acts as a temporary file that can store an arbitrary amount of data.

Once data generation is complete, the output will be directed to one or more output files. If no section_map specification is present, output will be directed to standard out. The sections will be output in the order specified in the DGL grammar. If a section_map specification is present, it controls the output process.

S1: section;
S2: section;
S3: section;
S4: section;
S5: section;

Figure 18. Section Declarations.

The section_map specification is a DGL statement, not a production. Only one section_map specification can be specified in a grammar. The section map specification is the keyword section_map followed by a list of sections. Figure 17 illustrates.

Section_Map:S2,S1,S1(abc.txt),S5(abc.txt),S4(data.txt),S3(echo.txt),S3(echo.txt);
Figure 19. The section map.

Figure 17 gives an example of an infile production. The infile production is used to input raw unformatted data one line at a time. Lines will be passed through the interpreter. Each reference of the form %{a} will cause one line to be read from the named file. The file name may contain non-terminals which are interpreted when the file is opened. If the file is not opened explicity, it will be opened automatically upon the first %a reference. The file name may contain non-terminals and will be reinterpreted if the file is closed and reopened.

a: **infile** "MyInput.txt";

Fig. 17. Infile productions.

7 Conclusion

DGL Version 2 should become an effective tool for testing interactive applications. We envision the data generation code being embedded in the application itself, with conditional compilation used to exclude it from released code. DGL is an on-going project, and we anticipate adding new features as the work progresses. The latest improvements have greatly enhanced its capabilities.

References

1. Maurer, P.: Generating test data with enhanced context free grammars. IEEE Softw. **7**(4), 50–56 (1990)
2. Maurer, P.: The design and implementation of a grammar-based data generator. Softw. Pract. Exper. **22**(3), 223–244 (1992)
3. Ahrens, J.H., Dieter, U.: Computer methods for sampling from Gamma, Beta, Poisson and Binomial distributions. Computing **12**, 223–246 (1974)
4. Aiello, W., Chung, F., Lu, L.: A random graph model for power law graphs. Exp. Math. **10**, 53–66 (2001)
5. Cheng, R.C.H.: The generation of gamma variables with non-integral shape parameter. J. Royal Stat. Soc. Ser. C (Appl. Stat.) **26**, 71–75 (1977)
6. Cheng, R.C.H.: Generating Beta variates with nonintegral shape parameters. Commun. ACM **21**, 317–322 (1978)
7. Marsaglia, G., Tsang, W.: The ziggurat method for generating random variables. J. Stat. Sci. **5**(8), 1–7 (2000)
8. Booth, T.L., Thompson, R.A.: Applying probability measures to abstract languages. In: IEEE Trans. Computers, pp. 442–450, May 1973
9. Maurer, P.: random database creation and use in a context-free grammar data generator. In: the 18th International Conference on Data Science (ICDATA 2022), July 2021
10. Maurer, P.: Massive generation of data with random variates. In: the 2021 Annual Modeling and Simulation Conference, July 2021

A Comparative Evaluation of Image Caption Synthesis Using Deep Neural Network

Sadia Nasrin Tisha[✉], Md Shahidur Rahaman, and Pablo Rivas

Department of Computer Science, School of Engineering and Computer Science,
Baylor University, Waco, TX, USA
{Sadia_Tisha1,MdShahidur_Rahaman1,Pablo_Rivas}@Baylor.edu

Abstract. Image caption generation is a crucial challenge in deep learning and natural language processing, involving identifying the context of an image and providing appropriate captions. In this study, we aimed to evaluate and compare the performance of two different model architectures using pre-trained CNN models for image classification and sequential LSTM models for caption generation. Specifically, we used RestNet50 and inceptionV3 CNN models with word2Vec and GloVe word embeddings, respectively, to generate captions. We evaluated the models based on two criteria: calculating the BLEU score for each generated caption and comparing the BLEU score with the inceptionResNetV2 state-of-the-art model. Our results showed that the second model architecture with inceptionV3 and GloVe-based model outperformed the first model and closely followed the benchmark BLEU score of the state-of-the-art model. Therefore, our study provides evidence that the choice of pre-trained CNN model and word embedding technique can significantly impact the performance of image caption generation, with the proposed architecture offering an accurate and efficient solution.

Keywords: Image caption generation · ResNet50 · InceptionV3 · LSTM · GloVe · Word2vec · BLEU score

1   Introduction

Generating image captions is a fundamental challenge in computer vision and natural language processing. It involves creating descriptions for images that express the objects and scenes present in them and how those objects and backgrounds relate to one another. Recent advancements in machine learning, particularly in convolutional and recurrent neural networks, have significantly improved the quality of caption generation [3,15]. The most successful models to date have been based on pre-trained convolutional neural networks, such as VGG-Net and inceptionResNetV2, and employ GloVe for word embedding [8].

However, selecting an appropriate configuration for generating accurate and efficient captions remains a significant challenge. This project aims to evaluate two different model architectures for image caption generation and provide

H. Han and E. Baker (Eds.): SDSC 2023, CCIS 2113, pp. 184–196, 2024.
https://doi.org/10.1007/978-3-031-61816-1_13

insight into which model is most appropriate to achieve state-of-the-art BLEU scores [10]. We compared the performance of two pre-trained CNN models, resNet50 and inceptionV3, for image classification, combined with long short-term memory (LSTM) sequential models as decoders. For the first model, we used RestNet50 CNN-based image feature extraction, word2vec word embedding, and LSTM to generate image captions. Finally, we employed the Inception-V3 model with GloVe word embedding and LSTM for sequential caption generation for the second model. We then evaluated both models based on their BLEU scores and compared the performance of the best-performing model with that of the state-of-the-art model, InceptionResNetV2.

The choice of an efficient model for generating image captions that balance execution time and accuracy is crucial. This study provides valuable insights into image caption generation's encoding and decoding phases and helps select the appropriate configuration for this task. With the availability of large classification datasets, such as COCO [4], Flickr [6], and Nocaps [1], and advanced training in deep neural networks, the generation of accurate and efficient captions for images has become more feasible. The results of this study demonstrate that the second model architecture, based on inceptionV3 and GloVe word embedding, is more effective in generating accurate and efficient image captions than the first model architecture.

The main contributions of our paper can be summarized as follows:

- Demonstration of the effectiveness of different image classification and word embedding techniques in image caption generation models.
- Identification of a more effective model, Model 2, which utilizes the inceptionV3 CNN model and GloVe word embedding technique to generate accurate and meaningful image captions.
- Findings that suggest the choice of image classification and word embedding techniques significantly impact the accuracy of image caption generation models.

2   Related Work

Recently, researchers extensively evaluated image caption generation using encoder-decoder-based models. Vinyals et al. [15] propose a neural and probabilistic framework that generates descriptions from images. In this study, the author encoded the variable length input into a fixed dimensional vector using an RNN model and obtained the desired output sentence by decoding the representation. The author uses arg max likelihood to increase the likelihood during training. Here, the author extracts the LSTM function and updates the memory block using a nonlinear function. The author employed CNN for image classification to feed a rich vector representation of the picture to RNN, which serves as a decoder. CNN is an innovative way to batch normalization. However, the researchers needed to provide which CNN model could best fit for encoding. Instead, the evaluation focused mainly on several datasets, e.g., Pascal [16], Flickr30k [6], COCO [4], and SBU [9].

Kesavan et al. [7] proposed a deep neural network to generate the caption with a pre-trained VGG-16 model. The researchers used CNN to create the thought vector for image extraction, which GRU uses as an RNN. The authors provide the state-of-the-art CNN configuration and generate the performance using the BLEU score. However, the model evaluation for the configuration is missing, as the aim was to achieve the state-of-the-art BLEU score from the image captions. In addition, Chen et al. [20] suggest a recurrent neural network to learn the bi-directional mapping between images and their sentence-based descriptions. Using a novel recurrent visual memory that automatically retains to recall long-term visual notions, they enable the development of innovative phrases given an image and utilize it to help in sentence generation and the reconstruction of visual features. These involve image retrieval, sentence production, and sentence retrieval. This paper presented state-of-the-art outcomes for creating innovative graphic descriptions for evaluation, where humans favor their automatically generated captions more frequently than 19.8% of the time when compared to captions that humans made. For techniques utilizing similar visual cues, performance on the picture and sentence retrieval tests is superior to or on par with state-of-the-art findings.

So we provided a comprehensive model analysis where the best-configured model can generate state-of-the-art captions, which will be measured by calculating BLEU scores.

3   Methodology

3.1   Dataset Description

We have used Flickr8K dataset [11]. The Flickr8k dataset is one of the standard datasets used in various image-related tasks. With 8,000 images paired with five different English captions that provide in-depth explanations of the key components and events, this dataset provides a collection of sentence-based image descriptions and searches. We added each image to five different English sentences that describe the image. We preprocessed the dataset by removing the noise so the model could detect the patterns easily in the text data. Here we cleaned the text's special characters, such as hashtags, punctuation, and numbers. The total dataset has a size of 1 GB. For training our model, we have divided our dataset into the training, testing, and validation sets, where 80% of images are in the training set, 10% in the testing set, and 10% in the validation set.

3.2   Model Description

In this project, we have generated a caption from an image using two model architectures. We have used convolutional neural networks (CNN) for image classification and a sequential LSTM model for developing captions. In the first architecture, we encoded the images using the CNN-based feature extraction architecture ResNet 50 and used word2vec embedding to obtain a vector representation for each corresponding sentence word. The LSTM layers would take partial captions-generated vectors as input and output the following word in the caption sequence. In the second architecture, we encoded the images using a pre-trained CNN model (inceptionV3) and utilized GloVe embedding to produce a vector representation for each corresponding text word. The LSTM layers would receive the partially captioned vectors and output the following word in the caption sequence. The next section is a description of our two models and their architecture:

Model 1: We have generated our model architecture with ResNet 50 CNN-based architecture for image classification in the first model. We retrieved the image's features in this instance just before the final classification layer. A 2048-bit vector is created by converting an additional dense layer. For word vectorization, we have used word2vec word embedding techniques. We tokenized 8253 unique words from the training dataset to define the vocabulary. As computers do not understand English words, we have represented them with numbers and mapped each vocabulary word with a unique index value. We encoded each word into a fixed-sized vector and defined each word as a number. Then we used the LSTM layers that take the partial caption-generated vectors as input and the image feature vectors as output, and they output the following word in the caption sequence for each test image. Figure 1 shows the total architecture of Model 1.

Figure 2 shows the model architecture where we used a $224 \times 224 \times 3$ dimensional image as the first input layer. For better classification, we used maxpooling for extracting the image feature from the vector and padding the vectors. ReLu activation is used in LSTM to decode the vectors with the softmax activation function. Here, we used the categorical cross-entropy loss function to calculate the loss, and for a 0.2 dropout rate, we got the minimum error in our training set. After successfully training the model, we test the model by providing test image data as input, which gives us the upcoming word in the caption sequence.

Model 2: In this model, we have used InceptionV3 CNN-based architecture for image classification and encoded the image into a feature vector. Then we used the vector for the input layer of the sequential LSTM model. InceptionV3 is a pre-trained model that extracts image features using three different sizes (e.g., $1 \times 1, 3 \times 3, 5 \times 5$) convolution and one max pooling, which classify the images with deeper layers. We utilized Glove embeddings for encoding, creating a vector representation of each sentence's words. GloVe stands for global vectors for word representation that generates word embeddings by aggregating a global word-word co-occurrence matrix from a corpus [14]. LSTM layers receive vector image features and vocabulary inputs and predict the test image's caption as an output. Figure 4 shows the total architecture of model 2.

Figure 3 shows the model architecture. Here, we have used $299 \times 299 \times 3$ dimensional image as the input layer. We have done the max pooling with a

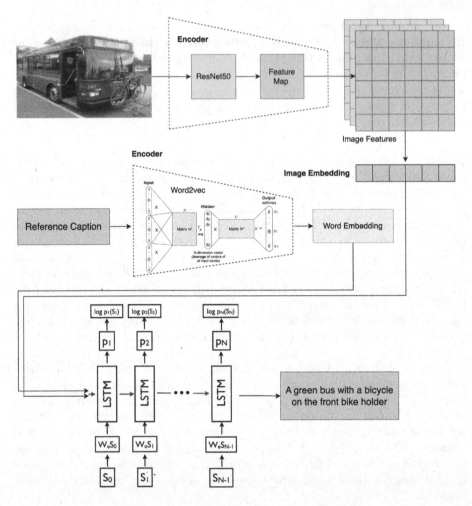

Fig. 1. Overall model architecture of Model 1.

2048 dense layer for extracting the image feature to a vector with batch normalization for better training. For word embedding, we have used Glove, with a total vocabulary size of 8763 with 256 dense layers and a 0.2 dropout rate. In addition, reLu activation and softmax activation functions are used in LSTM to decode the vectors, which improves the model error and training loss. Finally, after training, we put the model to the test by giving it test image data as input, and the model generates sequential sentences as captions.

3.3   Methods

Convolutional Neural Networks (CNN) for Image Classification: Convolutional neural networks are specialized deep neural networks that can process

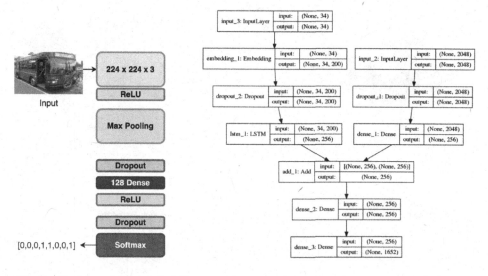

Fig. 2. Model architecture of Res Net50.

Fig. 3. Model architecture of model 2.

data in the form of a 2D matrix as input. It scans images from left to right and top to bottom to extract significant elements before combining them to classify them [17]. In addition, we have used CNN-based feature extraction architectures, ResNet 50, and Inception v3 architectures in this project.

1. **ResNet50:** The encoder step of image feature extraction using CNN has a specific expression ability, which is crucial in determining the quality of the image caption model. ResNet50 introduced a residual learning framework that is simple to optimize and has a low computing impact. We calculate the residual error to design and handle degradation and gradient problems, improving the network's performance as the depth grows. The model takes an image and produces a caption encoded as a sequence of $1 - k$ coded words.

$$y = y_1, y_2, y_3 \ldots y_c, y_i \in R^k$$

Here, k is the size of the dictionary, and c is the caption length. We use CNN, particularly ResNet50, to obtain set annotation vectors like the feature vectors. The extractor produces L-vectors, all of which are a D-dimensional representation of the corresponding part of an image.

2. **Inception v3:** In image classification tasks, the researchers commonly utilized Inception v3. The Inception module typically contains three different sizes of convolution and one maximum pooling. The channel is aggregated after the convolution process for the preceding layer's network output, and then nonlinear fusion is conducted [18]. Finally, the Inception v3 network structure uses a convolution kernel splitting method to divide large volume integrals into small convolutions.

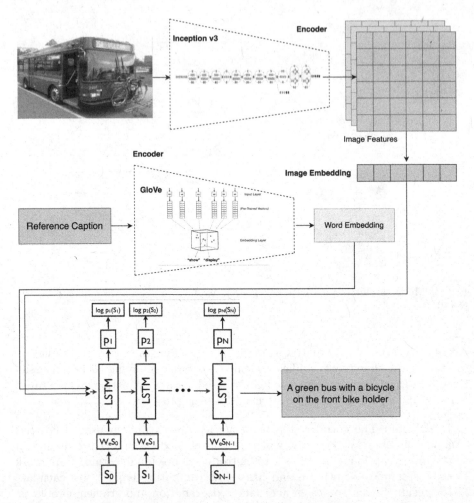

Fig. 4. Overall model architecture of model 2.

LSTM (Long Short-Term Memory): The Long Short-Term Memory Network (LSTMN) is an enhanced RNN (sequential network) that permits information to be stored indefinitely. LSTMs vary from standard feedforward neural networks in that they feature feedback connections. This trait allows LSTMs to process entire data sequences without having to handle each point separately instead of preserving necessary knowledge about primary data in the sequence to aid in processing incoming data points [19]. This project uses the LSTM language model to generate proper captions based on the input vector from the ResNet50 and inception v3 output.

$$f_t = \sigma(W_f[h_{t-1}, x_t] + b_f)$$

The output vector of the previous cell h_{t-1} with the new element of the sequence x_t are concatenated and passed as one vector through the layer with the softmax activation function.

4   Result Analysis

4.1   Model Result

For Model 1, we have trained the model with 300 epochs. Here Fig. 5 and 6 show after 150 epochs, the loss function is minimized to 0 where the training error is 0, and we got 85% accuracy for training. Finally, we generate the caption of test image data. The image on Table 1 top row shows the output caption of the image generated by Model 1, where the caption tells the correct sentence which describes the image.

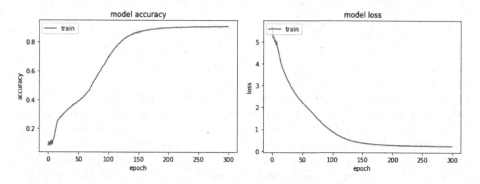

Fig. 5. The Accuracy graph of Model 1.   **Fig. 6.** The loss graph of Model 1.

In addition, for Model 2, we trained the model with 30 epochs, and Fig. 7 shows that the validation and training errors are minimized to 0. We got higher accuracy of 98% in both the training and validation sets. The image in the bottom row of Table 1, shows the output caption of the image generated by Model 2, which correctly identifies the image.

4.2   Model Evaluation

In this project, to evaluate the two models, first, we have calculated the BLEU score for each generated caption of the testing images and second, we compare the BLEU score with state-of-the-art model (e.g., inceptionResNetV2) results.

BLEU Score Calculation: We use the Bilingual Evaluation Understudy (BLEU) to evaluate the generated captions. The BLEU method is widely used to assess Natural Language Processing (NLP) systems that create language, particularly in natural language creation and machine translation [12]. The number in BLEU indicates the n-gram that the BLEU analyzes. This measure computes

Table 1. Captions generated by the two models

Model	Image	Generated caption
1		Two girls are lying upside down on a white bed.
2		Two football players are fighting over the ball.

the similarity score between produced and target text, which ranges from 0 to 1, with 1 indicating similarity and 0 indicating no resemblance [8]. In this project, for both Model 1 and Model 2, we have calculated the BLUE score for each generated caption for images with its reference captions from text data. Table 2 shows the BLEU score generated from Model 1, and the table also shows the BLUE score for Model 2. The table indicates that Model 2, with inception v3 image classification and GloVe word embedding model, gives a higher BLEU score than Model 1.

State-of-the-Art Model Evaluation: According to Keras application [13], inceptionResNetV2 is the most accurate image classification model. inception-ResNetV2 is another pre-trained model of convolutional neural architecture that builds on the Inception family of architectures but incorporates residual connections. Using our method, we utilized this pre-trained model and its image captions [2] and then computed the BLEU score for each caption. Finally, we compared the BLEU score with our calculated BLEU score for both models. Table 3 shows that, for this image, Model 2 gives a 0.6012 BLEU score, which is close to the inceptionResNetV2(state-of-the-art) model. On the other hand, Model 1 also shows a closer BLEU score than inceptionResNetV2.

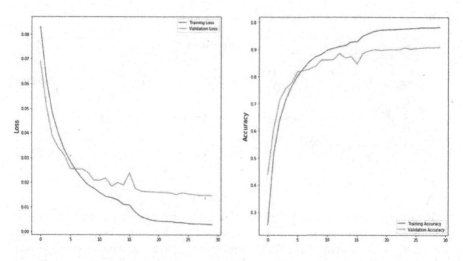

Fig. 7. The loss and accuracy graph of Model 2

Table 2. Captions generated by the two models

Image	Caption Model 1	BLEU	Caption Model 2	BLEU
	A boy in a life jacket is jacket is top on the side of a boat	0.44	Young boy in swim trunks is	0.66
	A boy players with a stringless racket in his backyard	0.40	Little girl in pink sweatshirt is holding racket	0.50

5   Discussion

The results of this project suggest that the combination of the inceptionV3 CNN model and GloVe word embedding technique used in Model 2 is more effective in generating accurate and meaningful captions for images than Model 1. The higher accuracy of Model 2 can be attributed to using a more advanced image classification model, which can identify more intricate features in the images, as well as a more sophisticated word embedding technique. These findings suggest that the choice of image classification and word embedding techniques can significantly impact the accuracy of image caption generation models.

Table 3. Comparative analysis of the model with its caption and BLEU score

Image	Model	Generated Caption	BLEU
	Inception-ResNetV2 (State of the art)	Young boy jumps on bed	0.623
	ResNet50 (Model 1)	A boy off a green, shirt jumps	0.549
	InceptionV3 (Model 2)	Boy jumps off bed	0.601

Moreover, comparing the calculated BLEU score with state-of-the-art model results indicates that the generated captions of Model 2 are closer to the benchmark solution, suggesting that this model can outperform other state-of-the-art models. However, it is essential to note that the results may vary depending on the dataset and image types used.

Another interesting finding from this project is that even though Model 1 has a lower accuracy and BLEU score than Model 2, the generated captions are still relatively close to the benchmark solution. This suggests that even a less advanced image caption generation model can still generate meaningful captions for images.

6   Conclusions

In conclusion, this project demonstrates the effectiveness of using different image classification and word embedding techniques in image caption generation models. Model 2, which utilizes the inceptionV3 CNN model and GloVe word embedding technique, outperforms Model 1 in terms of accuracy and BLEU score. The results also suggest that the generated captions of Model 2 are close to the benchmark solution, indicating that this model has the potential to outperform other state-of-the-art models.

However, there is still room for improvement in image caption generation models, and further research is needed to explore the use of other image classification and word embedding techniques. Nonetheless, this project contributes to the growing field of image caption generation and provides insights into the effectiveness of different techniques in generating accurate and meaningful captions for images.

References

1. Agrawal, H., et al.: Nocaps: novel object captioning at scale. In: Proceedings of the IEEE/CVF International Conference on Computer Vision, pp. 8948–8957 (2019)
2. Bhatia, Y., Bajpayee, A., Raghuvanshi, D., Mittal, H.: Image captioning using Google's inception-resnet-v2 and recurrent neural network. In: 2019 Twelfth International Conference on Contemporary Computing (IC3), pp. 1–6. IEEE (2019)
3. Chen, M., Ding, G., Zhao, S., Chen, H., Liu, Q., Han, J.: Reference based LSTM for image captioning. In: Thirty-First AAAI Conference on Artificial Intelligence (2017)
4. García, C.: MS-COCO-ES: Spanish coco captions (2019). https://www.kaggle.com/datasets/colmejano/mscocoes-spanish-coco-captions
5. Hossain, M.Z., Sohel, F., Shiratuddin, M.F., Laga, H.: A comprehensive survey of deep learning for image captioning. ACM Comput. Surv. (CsUR) **51**(6), 1–36 (2019)
6. HSANKESARA. Flickr image dataset flickr (2021). https://www.kaggle.com/datasets/hsankesara/flickr-image-dataset
7. Kesavan, V., Muley, V., Kolhekar, M.: Deep learning based automatic image caption generation. In: 2019 Global Conference for Advancement in Technology (GCAT), pp. 1–6. IEEE (2019)
8. Mahadi, M.R.S., Arifianto, A., Ramadhani, K.N.: Adaptive attention generation for Indonesian image captioning. In: 2020 8th International Conference on Information and Communication Technology (ICoICT), pp. 1–6. IEEE (2020)
9. Ordonez, V., Kulkarni, G., Berg, T.: Im2text: describing images using 1 million captioned photographs. In: Advances in Neural Information Processing Systems, vol. 24 (2011)
10. Papineni, K., Roukos, S., Ward, T., Zhu, W.-J.: Bleu: a method for automatic evaluation of machine translation. In: Proceedings of the 40th Annual Meeting of the Association for Computational Linguistics, pp. 311–318 (2002)
11. Rashtchian, C., Young, P., Hodosh, M., Hockenmaier, J.: Collecting image annotations using amazon's mechanical turk. In: Proceedings of the NAACL HLT 2010 Workshop on Creating Speech and Language Data with Amazon's Mechanical Turk, pp. 139–147 (2010)
12. Reiter, E.: A structured review of the validity of bleu. Comput. Linguist. **44**(3), 393–401 (2018)
13. Keras Team. Keras documentation: Keras applications
14. Tifrea, A., Bécigneul, G., Ganea, O.-E.: Poincaré glove: hyperbolic word embeddings. arXiv preprint arXiv:1810.06546 (2018)
15. Vinyals, O., Toshev, A., Bengio, S., Erhan, D.: Show and tell: a neural image caption generator. In: Proceedings of the IEEE Conference on Computer Vision and Pattern Recognition, pp. 3156–3164 (2015)
16. ZARAK. Pascal voc 2007 (2018). https://www.kaggle.com/datasets/zaraks/pascal-voc-2007
17. Chauhan, R., Ghanshala, K.K., Joshi, R.C.: Convolutional neural network (CNN) for image detection and recognition. In: 2018 First International Conference on Secure Cyber Computing and Communication (ICSCCC), pp. 278–282. IEEE (2018)
18. Dong, N., Zhao, L., Chun-Ho, W., Chang, J.-F.: Inception V3 based cervical cell classification combined with artificially extracted features. Appl. Soft Comput. **93**, 106311 (2020)

19. Yong, Yu., Si, X., Changhua, H., Zhang, J.: A review of recurrent neural networks: LSTM cells and network architectures. Neural Comput. **31**(7), 1235–1270 (2019)
20. Chen, X., Zitnick, C.L.: Learning a recurrent visual representation for image caption generation. arXiv preprint arXiv:1411.5654 (2014)

Parameter Estimation in Biochemical Models Using Marginal Probabilities

Kannon Hossain$^{(\boxtimes)}$ ⑩ and Roger B. Sidje$^{(\boxtimes)}$ ⑩

Department of Mathematics, The University of Alabama, Tuscaloosa, AL 35487, USA
khossain@crimson.ua.edu, roger.b.sidje@ua.edu

Abstract. Estimation of model parameters from experimental or synthetic data is an essential technique for working with stochastic models and is of increasing interest. We formulate the objective function through a fitting scheme based on a maximum likelihood estimator (MLE) that uses the marginal distribution of the species involved, which is a new way not attempted before. The quality of the method is evaluated for some example models, such as the Michaelis-Menten enzyme kinetics and mono-molecular reaction chain. Our numerical tests are performed with both local and global optimization schemes. It is shown that the method performs well compared to existing approaches.

Keywords: Parameter estimation · Stochastic model · Finite state projection · Maximum likelihood estimator

1 Introduction

The chemical master equation (CME) provides a mathematical framework to model stochastic biochemical reaction networks [1,2]. The CME describes the evolution of the networks over time through their various configurations, i.e., the population counts of the chemical species involved. However, even using numerical approximations, the CME is often not directly solvable because of its large and potentially infinite dimension. Owing to the finite state projection (FSP) [3], which is an advancement in the numerical treatment of the CME, a reasonable approximation to the solution can be achieved by projecting the CME to a reduced state space. Nonetheless, even for modest biochemical networks, the FSP method continues to produce high-dimensional models that are resource-intensive to run. For this reason, ongoing works remain on how well the CME can be simulated, and newer techniques have been created that can be utilized with even more complex networks [4–7].

In systems biology, parameter estimation plays a crucial role in model analysis. To avoid costly trial-and-error lab experiments, systems biology relies heavily on computational modeling to shed light on extremely complex biological processes. Typically, parameter estimation is an iterative procedure, with implementations mostly based on deterministic models [8–11] than on the much more challenging stochastic models [12–14]. But deterministic strategies assume mathematical formulations not purposely designed to capture the randomness that occurs when species only exist in a few copies. In these situations, the corresponding stochastic modeling may be a better way to go, but the computation is very time-consuming [15, 16]. In any case, every approach for estimating parameters, whether deterministic or stochastic, has its own set of advantages and disadvantages. Discussions pertaining to these aspects can be found in the literature [17–20].

A common modeling problem involves how to estimate a joint probability distribution for a dataset. Given observed data, the maximum likelihood estimation method finds the parameters of an assumed probability distribution. To achieve this, a probability function is maximized to make the observed data most likely to fit the assumed statistical model. The estimate with the highest possible likelihood is called the maximum likelihood estimate, and it refers to the point in the parameter space where the likelihood function is maximum. We represent a set of data as a random sample from an unknown joint probability distribution stated in terms of a set of parameters.

In this work, we apply the idea of the maximum likelihood estimator (MLE), which has been viewed as a natural method given the probabilistic character of stochastic models [21]. This, in turn, ultimately boils down to an optimization problem. But further complications arise because the objective function may have multiple modes, together with confounded parameters and identifiability issues in the model. The inherent complexity of modeling biological systems leads to optimizing nonlinear and non-convex functions with possibly many, locally optimal solutions. As a result, classic gradient-based approaches for estimating parameters may be trapped into local optima and suffer to find the best fit. One solution to overcome this issue could be to use a derivative-free or global algorithm. While several derivative-free and global optimization [22] strategies have been compared, no one optimization strategy consistently outperforms the others in all test cases. About more than 100 potential approaches [23] and their variants make it challenging for non-specialists and professionals to appraise competing alternatives and their restrictions. In light of this, it is essential to examine various optimization algorithms in the scope of parameter fitting using maximum likelihood, which is one of the most used optimization strategies. Here a comparison of the estimated parameters using the state transition probability and marginal distribution of each species is presented with two different optimization schemes to calibrate the model parameters.

The rest of the paper is organized as follows: In Sect. 2, we summarize the CME, the structure of the likelihood function, parameter fitting technique, and optimization algorithms. Sections 3 and 4 show numerical experiments, test results, discussions, and sensitivity effects. Lastly, concluding remarks are drawn in Sect. 5.

2   Methods

2.1   CME and FSP

Consider a biochemical process of N chemical species interacting via M reactions. The state of the system at any time t is a vector $\mathbf{x} = (x_1, \cdots, x_N)^T$, where x_i is the population count of the i-th species, $1 \le i \le N$. Let $P(\mathbf{x}, t)$ be the probability that the system is at state $\mathbf{x}$ at time t. The chemical master equation (CME) [2] states that

$$\frac{dP(\mathbf{x}, t)}{dt} = \sum_{k=1}^{M} \alpha_k (\mathbf{x} - \boldsymbol{\nu}_k) P(\mathbf{x} - \boldsymbol{\nu}_k, t) - \sum_{k=1}^{M} \alpha_k (\mathbf{x}) P(\mathbf{x}, t), \qquad (1)$$

where, for the k-th reaction, $1 \le k \le M$, the $\boldsymbol{\nu}_k$ and α_k are, respectively, the stoichiometric vector and propensity function.

Because of the "curse of the dimensionality", the size of CME can be extremely large or theoretically infinite. The FSP [3] makes a truncation with an analytical bound on the error of the probability distribution. As the number of states taken into account in the FSP is increased, this error bound is decreased and the probability of any given state of the system is more accurate. Keeping n states, we can write (1) as a system of linear ordinary differential equations (ODEs)

$$\dot{\mathbf{p}}(t) = \mathbf{A} \cdot \mathbf{p}(t), \quad t \in [0, t_f], \qquad (2)$$

where the transition rate matrix $\mathbf{A} = [a_{ij}] \in \mathbb{R}^{n \times n}$ is defined as

$$a_{ij} = \begin{cases} -\sum_{k=1}^{M} \alpha_k (\mathbf{x}_j), & \text{if } i = j, \\ \alpha_k (\mathbf{x}_j), & \text{if } \mathbf{x}_i = \mathbf{x}_j + \boldsymbol{\nu}_k, \\ 0, & \text{otherwise.} \end{cases}$$

If the initial state of the system at time $t = 0$ is known, the probability vector of the system at the time point t_f is given by

$$\mathbf{p}(t_f) = \exp(t_f \mathbf{A}) \mathbf{p}(0) \qquad (3)$$

where the exponential matrix is defined as

$$\exp(t_f \mathbf{A}) = \sum_{m=0}^{\infty} \frac{(t_f \mathbf{A})^m}{m!}. \qquad (4)$$

2.2   Parameter Estimation Technique

There have been several MLE investigations [24–27] that used Monte Carlo or the stochastic simulation algorithm (SSA) to solve the CME by drawing random trajectories of the system and utilizing the generated frequencies to indirectly approximate the actual probability distributions allowing them to circumvent the "curse of dimensionality". On the other hand, FSP solves the CME directly

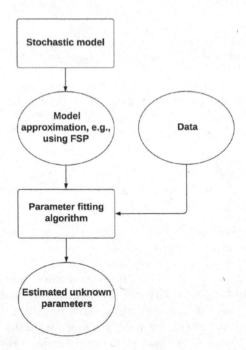

Fig. 1. A flow-chart for parameter estimation in biochemical models.

by restricting each species number and thus reducing the total state space. Our approach here is to first utilize the FSP to approximately solve the CME that provides the state transition probability vector, and later on, use it to approximate the marginal distribution of each species. Since CME is a high-dimensional or infinite-dimensional system, estimating its parameters is a challenging task. To cope with the so-called "curse of dimensionality" recent research on parameter estimation [28] shows that using bayesian inference tasks for the chemical master equation in the tensor-train format can be very effective (Fig. 1).

When setting up the CME in (1) or (2), there is an implicit dependency on the constant reaction rates embedded in the system. To first make it explicit that the behavior of the system depends on some parameters $\boldsymbol{\theta} = (c_1, \ldots, c_M)^T$ that are to be estimated, we now write

$$\dot{\mathbf{p}} = \mathbf{A}(\boldsymbol{\theta}) \cdot \mathbf{p}, \quad t \in [0, t_f]. \tag{5}$$

for which the solution is the probability vector $\mathbf{p}(t, \boldsymbol{\theta}) = (p_1(t, \boldsymbol{\theta}), \ldots, p_n(t, \boldsymbol{\theta}))^T$ where $p_\ell(t, \boldsymbol{\theta}) = P(\mathbf{x}_\ell, t, \boldsymbol{\theta})$ is the probability of finding the system in state $\mathbf{x}_\ell$ at time t, with the $\mathbf{x}_1, \ldots, \mathbf{x}_n$ being the states retained by the FSP.

Given the observed data as input parameter estimation uses numerical techniques to calculate a parameter that calibrates the model to the data. In this work, 100 SSA realizations were used to generate a unique data set for testing purposes. For the data set $\mathcal{D}$ and at fixed time $t = t_f$, the likelihood function $\mathcal{L}_{\mathcal{D}}^{\text{FSP}}(\boldsymbol{\theta})$ is obtained as the product of the transition probabilities

$$\mathcal{L}_{\mathcal{D}}^{\text{FSP}}(\boldsymbol{\theta}) = \prod_{\ell} P(\mathbf{x}_\ell, t, \boldsymbol{\theta}) \tag{6}$$

and we restrict the running index ℓ in (6) to plausible states determined by the data set $\mathcal{D}$. The log-likelihood function allows us to re-frame the problem to improve numerical stability, with the goal of determining the parameter set $\boldsymbol{\theta}_{\text{Fit}}$ which will maximize (6). Therefore the FSP-based MLE problem is:

$$\boldsymbol{\theta}_{\text{Fit}} = \arg\max_{\theta}(\mathcal{L}_{\mathcal{D}}^{\text{FSP}}(\boldsymbol{\theta})) \tag{7}$$

$$= \arg\max_{\theta}(\log(\mathcal{L}_{\mathcal{D}}^{\text{FSP}}(\boldsymbol{\theta}))) \tag{8}$$

$$= \arg\max_{\theta}\left(\sum_{\ell} \log(P(\mathbf{x}_\ell, t, \boldsymbol{\theta}))\right). \tag{9}$$

Each evaluation of the objective function with a different $\boldsymbol{\theta}$ implies solving the CME to retrieve the $P(\mathbf{x}_\ell, t, \boldsymbol{\theta})$, and this is why efficient solution techniques are critical. A new approach that we have proposed and that we shall investigate here is to replace the state transition probabilities with the marginal distribution of each species. Either way, maximizing the likelihood, or correspondingly minimizing the negative log-likelihood, yields maximum-likelihood parameter estimates. We have employed one derivative-free local optimization scheme *fminsearch* and one global optimization schemes *Multistart* in MATLAB [29] to conduct our numerical test.

2.3   Optimization Algorithms

Since the inherent nonlinearity of biological systems results in a non-convex non-linear programming problem (NLP), most classic nonlinear algorithms utilizing gradient approaches are in danger of being stuck at a local optimum. Because local search algorithms converge quickly, the most traditional and straightforward approach to overcome the non-convexity of many optimization problems is to repeatedly change the initial conditions in the hope of getting the best estimates. In some cases, there is theoretical proof of convergence to the optimum for local search techniques if they are started sufficiently near that optimum.

The MATLAB routine *fminsreach* that we use in this work is a derivative-free [30] local optimization routine meant for solving unconstrained multivariable nonlinear optimization problems. To avoid the search process from getting stuck at a local optimum, it is recommended to use global optimizers, most of which are stochastic in nature. Various categories of stochastic techniques may be used for global optimization. Examples include simulated annealing, colony optimization, particle swarm optimization, and many more. In our biological models, we have employed one global solver, *Multistart* in MATLAB. The solver, *Multistart* launches many local searches from several starting points in the parameter space and selects the best result within the basin of attraction. Related work on this global estimation approach [31] shows that it can successfully estimate the parameters of the biological models.

3   Numerical Experiment with a Mono-molecular Reaction Chain Model

We first consider a simple mono-molecular reaction chain model [32] where the system consists of two reactions, R_1 and R_2; and three species A, B, and C. Through R_1, one individual A transforms into one individual B, and similarly through R_2, B transforms into C. The scalars c_1 and c_2 represent the stochastic rate constants or parameters of the model, with true values $\theta_{true} = (0.04, 0.11)$ stated in [32]. Table 1 lists the propensity functions and the state change vectors.

Table 1. Mono-molecular reaction chain model

	Reaction	Propensity	State change vector
R_1:	$A \xrightarrow{c_1} B$	$\alpha_1 = c_1[A]$	$\nu_1(A, B, C) = (-1, 1, 0)$
R_2:	$B \xrightarrow{c_2} C$	$\alpha_2 = c_2[B]$	$\nu_2(A, B, C) = (0, -1, 1)$

Since biochemical reaction models can be modeled with both deterministic and stochastic approaches, it is of interest to contrast the two. Figure 2 shows the simulations via reaction rate equation (RREs) and stochastic simulation algorithm (SSA) within the time frame of $t = 50$ using the initial conditions $([A], [B], [C]) = (30, 20, 0)$ and parameters $\theta_{\mathbf{true}} = (0.04, 0.11)$. Here we clearly see a similarity between the two outcomes.

3.1   Test Results

The efficiency of the optimization techniques is sensitive to the nature of the problems at hand, with no single optimization strategy is guaranteed to be the most effective in each and every scenario. The true parameters are typically unknown in advance in real-world applications. To get the best possible outcome from a local optimization method, one strategy would be to trial different initial guesses at random from a given range and select the best possible final estimation. The good thing about the derivative-free local optimizer *fminsreach* is that it can often handle discontinuity. Table 2 is showing the test results of the local optimization algorithm and Table 3 is showing the test results of the global optimization algorithm at time $t = 45$. The test results are discussed in the next section.

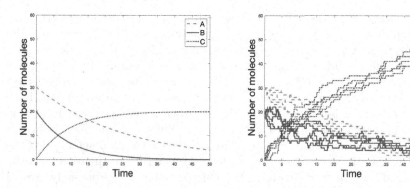

Fig. 2. Simulations of the mono-molecular reaction chain model via RREs (left) and five SSA realizations (right).

Table 2. Results of the local optimization algorithm

	ETP		EMD	
	c_1	c_2	c_1	c_2
True parameter	0.04	0.11	0.04	0.11
Initial guess	0.1	0.5	0.1	0.5
Test 1	0.0451	0.1135	0.0453	0.1112
Relative error	12.75%	3.18%	13.24%	1.09%
Initial guess	0.009	0.001	0.009	0.001
Test 2	0.0451	0.1134	0.0453	0.1113
Relative error	12.75%	3.09%	13.24%	1.18%

Table 3. Result of the global optimization algorithm

	ETP		EMD	
	c_1	c_2	c_1	c_2
True parameter	0.04	0.11	0.04	0.11
Initial guess	0.9	0.02	0.9	0.02
MultiStart	0.0459	0.1138	0.0457	0.1164
Relative error	14.75%	3.45%	14.24%	5.81%

3.2   Discussions

Comparison of ETP and EMD. It is well known that sometimes the local optimizer performs poorly if we choose an initial guess that is far from the correct parameters. To see how the initial guess changes the test results, we pick two sets of different initial conditions for our local optimizer, which are clearly shown in Table 2. For Test 1, we have seen that the initial condition is chosen close to the true parameter, and as a result, we obtained good estimations both in ETP and EMD. It is noticeable that the relative error compared with the true parameter is reasonable both in ETP and EMD. Since our contribution in this paper is to see whether or not it is possible to estimate the parameters using the marginal distribution of each species, we have seen that the estimation from the EMD is good enough to compare to the true parameters. Surprisingly, Test 2 gives us a good estimation for both ETP and EMD, even though we have chosen initial conditions far from the true parameters. The important observation here is that the differences between the estimation from ETP and EMD are relatively small, as evident in Table 2. The log-likelihoods resulting from the product of the state transition probabilities and the marginal distribution of each species, made the objective function relatively small, which stopped it from iterating further. To overcome these issues, we have adjusted the maximum function evaluation *MaxFunEvals* to 3000 and maximum iteration *MaxIter* to 3000 where we keep *TolX, TolFun* as 10^{-4} while we conduct our test in MATLAB.

The objective of the algorithms used in global optimization is to locate the maximum value throughout the entire range. *Multistart* runs a local solver *fmincon* from each set of start points to find the global maximum. In our optimization routine, we have chosen 20 instances, which means the solver attempts to find multiple local solutions to a problem by starting from 20 various points. Table 3 shows that with this algorithm the estimation of the parameters in both ETP and EMD is well inferred. The relative error is small compared to the true parameter. The *FunctionTolerance* and *XTolerance* was 10^{-6} when we performed this test. It is important that a range is given for each parameter in order to ensure that the values are reasonable from a biological standpoint. If the estimation tasks are ill-conditioned and multi-modal, then exploring the large parameter space can lead to a wrong estimation [33]. For this problem, we set the lower and upper bounds as [0, 0] and [1, 1] respectively.

It is necessary to show the fitting scheme as well as the comparative analysis graphically when we estimate the unknown parameters for our model. Here, Fig. 3 displays the fitting scheme for the state transition probability whereas Fig. 4 displays the fitting scheme for the marginal distribution of each species individually.

Sensitivity Effect. The dependency of the behavior of the system on the parameters that influence the dynamics of the process can be studied using sensitivity analysis [34]. Having a high sensitivity to a parameter means that even tiny changes in that parameter can have a significant impact on the system's performance, and vice versa. In real-life applications, the true likelihoods

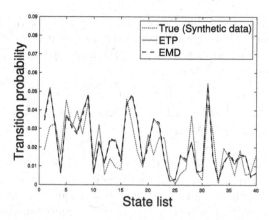

Fig. 3. Comparison of the state transition probabilities between the result from True (Synthetic data) and Test 1 that is fitted.

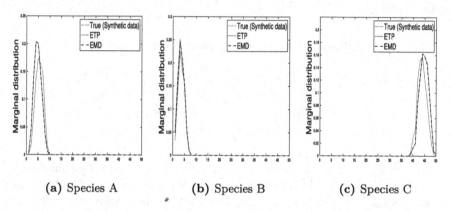

(a) Species A   (b) Species B   (c) Species C

Fig. 4. A comparison of the marginal probability distribution with the true and estimated (Test 1) parameter values of each species of the mono-molecular reaction chain model

Table 4. Comparison of the likelihood values.

	Likelihood	
	ETP	EMD
True (Synthetic data)	−170.44	−88.04
Test 1	−164.19	−84.54
Test 2	−164.24	−84.52
Multistart	−164.41	−84.41

are not known in advance. Table 4 shows the comparison of the likelihoods that are calculated with ETP and EMD. We have seen that the difference between the likelihood values between ETP and EMD varies much, indicating that the parameters are sensitive to our model.

4   Numerical Experiment with the Michaelis-Menten Enzyme Kinetics Model

The second example we consider is the Michaelis-Menten enzyme kinetics [35]. The biochemical system consists of the species: enzyme E, which catalyzes the reaction of substrates S into products P by forming intermediate enzyme-substrate complexes ES. The scalars c_1, c_2, and c_3 represent the stochastic rate constants or parameters of the model with true values $\theta_{true} = (1.0, 1.0, 0.1)$ stated in [35]. Table 5 lists the propensity functions and the state change vectors for this model.

Table 5. Michaelis-Menten enzyme kinetics model

	Reaction	Propensities	State change vector
R_1:	$S + E \xrightarrow{c_1} ES$	$\alpha_1 = c_1[S][E]$	$\nu_1(S, E, ES, P) = (-1, -1, 1, 0)$
R_1:	$ES \xrightarrow{c_2} E + S$	$\alpha_2 = c_2[ES]$	$\nu_2(S, E, ES, P) = (1, 1, -1, 0)$
R_1:	$ES \xrightarrow{c_3} P + E$	$\alpha_3 = c_3[ES]$	$\nu_3(S, E, ES, P) = (0, 1, -1, 1)$

Figure 5 shows the simulations via RREs and SSA realizations within the time frame of $t = 25$ using the initial conditions $([E], [S], [ES], [P]) = (30, 20, 0, 0)$ and parameters $\theta_{true} = (1.0, 1.0, 0.1)$ which depicts that the graphs are similar.

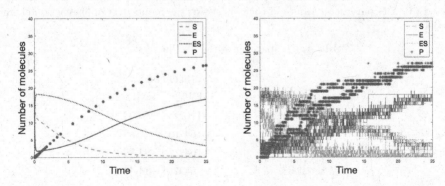

Fig. 5. Simulations of the michaelis-menten enzyme kinetics model via RREs (left) and five SSA realizations (right).

4.1 Test Results

Table 2 is showing the test results of the local optimization algorithm and Table 3 is showing the test results of the global optimization algorithm at time $t = 13$. The test results are discussed in the next section.

Table 6. Results of the local optimization algorithm

	ETP			EMD		
	c_1	c_2	c_3	c_1	c_2	c_3
True parameter	1.0	1.0	0.1	1.0	1.0	0.1
Initial guess	0.8	0.7	0.9	0.8	0.7	0.9
Test 1	0.9336	1.09	0.0768	0.9703	1.04	0.0813
Relative error	6.64%	9.0%	23.20%	2.96%	4.0%	18.70%
Initial guess	0.1	0.1	0.9	0.1	0.1	0.9
Test 2	0.1384	0.1397	0.0794	0.1450	0.1364	0.0842
Relative error	86.16%	86.03%	20.60%	85.5%	86.36%	15.80%

Table 7. Result of the global optimization algorithm

	ETP			EMD		
	c_1	c_2	c_3	c_1	c_2	c_3
True parameter	1.0	1.0	0.1	1.0	1.0	0.1
Initial guess	0.03	0.05	0.05	1.3	0.05	0.05
Multistart	1.01	1.19	0.0768	1.08	1.20	0.0815
Relative error	1.0%	18.99%	23.20%	8.0%	19.99%	18.50%

4.2 Discussions

Comparison of ETP and EMD. For this model, we have tested two different initial guesses for the local optimizer with the time points $t = 13$ to see how the estimations vary. All of the estimation results can be found in Table 6. Test 1 shows that the inferences of all three parameters, both from ETP and EMD, are good enough compared to the true parameters. The third parameter from ETP and EMD has a high relative error. But we can clearly see that the estimation using the marginal distribution is better than the existing method. To see how initial guesses really affect the optimization process in the local solver, for Test 2, we pick our initial guess far away from the true parameter. As a result, none of the estimated parameters from both ETP and ETD inferred well compared to the true parameters as they have high relative errors. We keep the same setup

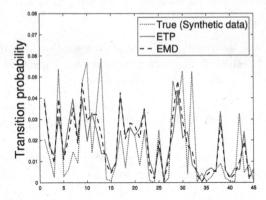

Fig. 6. Comparison of the state transition probabilities between the result from True (Synthetic data) and Test 1 that is fitted.

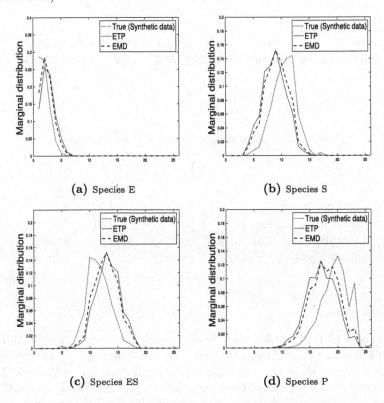

Fig. 7. A comparison of the marginal probability distribution with the true and estimated (Test 1) parameter values of each species of the Michaelis-Menten enzyme kinetics

Table 8. Comparison of the likelihood values

	Likelihood	
	ETP	EMD
True (Synthetic data)	−228.13	−192.34
Test 1	−199.30	−175.12
Test 2	−199.77	−175.90
Multistart	−199.17	−175.09

for *MaxFunEvals*, *MaxIter*, *TolX*, and *TolFun* that we had in the previous model to conduct the local optimization test.

In this case, for the *Multistart* method, we have chosen 20 instances, which means the solver attempts to find multiple local solutions to a problem by starting from 20 various points. Table 7 shows the test results of this solver. Again, we see that all of the parameters from both ETP and EMD are well estimated compared to the true parameters. We set the lower and upper bounds as [0, 0, 0] and [1.2, 1.2, 0.5], respectively, where the *FunctionTolerance* and *XTolerance* were 10^{-6}.

For this model, Fig. 6 displays the fitting scheme for the state transition probability whereas Fig. 7 displays the fitting scheme for the marginal distribution of each species individually.

Sensitivity Effect. Table 8 shows the comparison of the likelihoods that are calculated with ETP and EMD. We have seen that the difference between the likelihood values varies not too much, indicating that the parameters are not much sensitive to our model.

5   Conclusion

For biological processes that are inherently based on probabilities, the application of mathematical modeling relies on using stochastic models. In this work, we have shown that by solving the CME with the FSP, it is possible to estimate the parameters of a model, not only through state transition probability but also using the marginal distribution of each species. Moreover, we have found that the difference between the estimated parameters from ETP and EMD is relatively small. Given the significance of the parameter estimation problem, further investigations might require more extensive comparisons incorporating more biological models and more optimization algorithms.

References

1. Goutsias, J., Jenkinson, G.: Markovian dynamics on complex reaction networks. Phys. Rep. **529**(2), 199–264 (2013)
2. Gillespie, D.T.: A rigorous derivation of the chemical master equation. Physica A Stat. Mech. Appl. **188**(1–3), 404–425 (1992)
3. Munsky, B., Khammash, M.: The finite state projection algorithm for the solution of the chemical master equation. J. Chem. Phys. **124**(4), 044104 (2006)
4. Hegland, M., Garcke, J.: On the numerical solution of the chemical master equation with sums of rank one tensors. Anziam J. **52**, C628–C643 (2010)
5. Kazeev, V., Khammash, M., Nip, M., Schwab, C.: Direct solution of the chemical master equation using quantized tensor trains. PLoS Comput. Biol. **10**(3), e1003359 (2014)
6. Dolgov, S., Khoromskij, B.: Simultaneous state-time approximation of the chemical master equation using tensor product formats. Numer. Linear Algebra Appl. **22**(2), 197–219 (2015)
7. Oseledets, I.V.: Tensor-train decomposition. SIAM J. Sci. Comput. **33**(5), 2295–2317 (2011)
8. Moles, C.G., Mendes, P., Banga, J.R.: Parameter estimation in biochemical pathways: a comparison of global optimization methods. Genome Res. **13**(11), 2467–2474 (2003)
9. Ortiz, A.R., Banks, H.T., Castillo-Chavez, C., Chowell, G., Wang, X.: A deterministic methodology for estimation of parameters in dynamic Markov chain models. J. Biol. Syst. **19**(01), 71–100 (2011)
10. Miró, A., Pozo, C., Guillén-Gosálbez, G., Egea, J.A., Jiménez, L.: Deterministic global optimization algorithm based on the outer approximation for the parameter estimation of nonlinear dynamic biological systems. BMC Bioinform. **13**(1), 1–12 (2012)
11. Lin, Y., Stadtherr, M.A.: Deterministic global optimization for parameter estimation of dynamic systems. Ind. Eng. Chem. Res. **45**(25), 8438–8448 (2006)
12. Wang, Y.-C., Chen, B.-S.: Integrated cellular network of transcription regulations and protein-protein interactions. BMC Syst. Biol. **4**(1), 1–17 (2010)
13. Ashyraliyev, M., Fomekong-Nanfack, Y., Kaandorp, J.A., Blom, J.G.: Systems biology: parameter estimation for biochemical models. FEBS J. **276**(4), 886–902 (2009)
14. Villaverde, A.F., Fröhlich, F., Weindl, D., Hasenauer, J., Banga, J.R.: Benchmarking optimization methods for parameter estimation in large kinetic models. Bioinformatics **35**(5), 830–838 (2019)
15. Zimmer, C., Sahle, S.: Parameter estimation for stochastic models of biochemical reactions. J. Comput. Sci. Syst. Biol. **6**, 011–021 (2012)
16. Dinh, K.N., Sidje, R.B.: An application of the Krylov-FSP-SSA method to parameter fitting with maximum likelihood. Phys. Biol. **14**(6), 065001 (2017)
17. Kazeroonian, A., Hasenauer, J., Theis, F.: Parameter estimation for stochastic biochemical processes: a comparison of moment equation and finite state projection. In: The 10th International Workshop on Computational Systems Biology, WCSB 2013, Tampere, Finland, 10–12 June, p. 67 (2013)
18. Zimmer, C., Sahle, S.: Comparison of approaches for parameter estimation on stochastic models: generic least squares versus specialized approaches. Comput. Biol. Chem. **61**, 75–85 (2016)

19. Gupta, A., Rawlings, J.B.: Comparison of parameter estimation methods in stochastic chemical kinetic models: examples in systems biology. AIChE J. **60**(4), 1253–1268 (2014)

20. Simoni, G., Vo, H.T., Priami, C., Marchetti, L.: A comparison of deterministic and stochastic approaches for sensitivity analysis in computational systems biology. Brief. Bioinform. **21**(2), 527–540 (2020)

21. Wilkinson, D.J.: Stochastic modeling for quantitative description of heterogeneous biological systems. Nat. Rev. Genet. **10**(2), 122–133 (2009)

22. Rios, L.M., Sahinidis, N.V.: Derivative-free optimization: a review of algorithms and comparison of software implementations. J. Global Optim. **56**, 1247–1293 (2013)

23. Chou, I.-C., Voit, E.O.: Recent developments in parameter estimation and structure identification of biochemical and genomic systems. Math. Biosci. **219**(2), 57–83 (2009)

24. Wang, Y., Christley, S., Mjolsness, E., Xie, X.: Parameter inference for discretely observed stochastic kinetic models using stochastic gradient descent. BMC Syst. Biol. **4**(1), 1–16 (2010)

25. Daigle, B.J., Roh, M.K., Petzold, L.R., Niemi, J.: Accelerated maximum likelihood parameter estimation for stochastic biochemical systems. BMC Bioinform. **13**(1), 1–18 (2012)

26. Poovathingal, S.K., Gunawan, R.: Global parameter estimation methods for stochastic biochemical systems. BMC Bioinform. **11**(1), 1–12 (2010)

27. Tian, T., Songlin, X., Gao, J., Burrage, K.: Simulated maximum likelihood method for estimating kinetic rates in gene expression. Bioinformatics **23**(1), 84–91 (2007)

28. Ion, I.G., Wildner, C., Loukrezis, D., Koeppl, H., De Gersem, H.: Tensor-train approximation of the chemical master equation and its application for parameter inference. J. Chem. Phys. **155**(3), 034102 (2021)

29. MATLAB. version 7.10.0 (R2022a). The MathWorks Inc., Natick, Massachusetts (2022)

30. Lagarias, J.C., Reeds, J.A., Wright, M.H., Wright, P.E.: Convergence properties of the Nelder-Mead simplex method in low dimensions. SIAM J. Optim. **9**(1), 112–147 (1998)

31. Fröhlich, F., Kaltenbacher, B., Theis, F.J., Hasenauer, J.: Scalable parameter estimation for genome-scale biochemical reaction networks. PLoS Comput. Biol. **13**(1), e1005331 (2017)

32. Gupta, A.: Parameter estimation in deterministic and stochastic models of biological systems. Ph.D. thesis, The University of Wisconsin-Madison (2013)

33. Liu, P.-K., Wang, F.-S.: Hybrid differential evolution with a geometric mean mutation in parameter estimation of bioreaction systems with large parameter search space. Comput. Chem. Eng. **33**(11), 1851–1860 (2009)

34. Gunawan, R., Cao, Y., Petzold, L., Doyle, F.J.: Sensitivity analysis of discrete stochastic systems. Biophys. J . **88**(4), 2530–2540 (2005)

35. Goutsias, J.: Quasiequilibrium approximation of fast reaction kinetics in stochastic biochemical systems. J. Chem. Phys. **122**(18), 184102 (2005)

Detecting Microservice Anti-patterns Using Interactive Service Call Graphs: Effort Assessment

Austin Huizinga[1] , Garrett Parker[1] , Amr S. Abdelfattah[1] ,
Xiaozhou Li[2,3] , Tomas Cerny[4(✉)] , and Davide Taibi[2,3]

[1] Baylor University, Waco, TX 76706, USA
[2] Tampere University, Tampere, Finland
[3] Oulu University, Oulu, Finland
[4] SIE, University of Arizona, Tucson, AZ, USA
tcerny@arizona.edu

Abstract. Together with the increasing adoption of microservices, detecting microservice anti-patterns has become a crucial practice. However, the number of tools supporting effective anti-pattern detection is limited. Though involving the human in the loop is useful, it is time-consuming and lacks the accuracy necessary to complete such a task. For such a purpose, we consider visualizing the microservice system architecture using the service view, specifically the service call graph. In this paper, we present a framework to visualize service call graphs in an interactive 3D node-edge model. Utilizing an intermediate representation of the microservice system architecture, we create our interactive model to allow for quicker, more accurate detection.

Keywords: Microservices · Visualization · Smells detection · anti-patterns

1 Introduction

Microservices architecture is the mainstream direction that provides scalability, robustness, and decentralization. The system is broken down into autonomous microservices developed by separate teams. While software architects prescribe the microservice system design, they have no mechanism to assess the implemented design in the solution. Microservices evolve decentrally in many places at the same time, which can lead to ripple effects or deteriorate system quality [3]. Additionally, development teams are often pushed by their managers to develop new features and correct bugs which are visible to them, putting less emphasis on maintaining system design quality. This is partly because the quality is less visible to or has less direct value to managers.

One established way to maintain good system design properties is to avoid anti-patterns [11]. Yet, for microservices, these are more challenging to identify, given the decentralized environments where only limited prototype tools exist.

The goal of our paper is to support quality assurance in microservices by improving the effectiveness of anti-pattern detection using service call graphs.

H. Han and E. Baker (Eds.): SDSC 2023, CCIS 2113, pp. 212–227, 2024.
https://doi.org/10.1007/978-3-031-61816-1_15

We aim to investigate whether the automation of anti-pattern detection can help developers compared to manual anti-pattern detection. In comprehending the accuracy of manual detection of anti-patterns from the service view perspective, we conducted a case study involving junior developers. We focus on two common anti-patterns that can be shown from the service view: Cyclic Dependencies [12] and the knot [8], as a proof-of-concept. For this purpose, we developed a visualization framework for representing the service view through service call graphs. The framework uses an interactive 3D environment with a node-edge model for the service view. Compared to other popular tracing tools with similar system visualization features (e.g., Jaeger https://www.jaegertracing.io), our framework is also capable of anti-pattern highlighting, emphasizing potential system problems to developers.

The results of our study show that manual detection of these anti-patterns yields inaccuracies and a high time cost. Even developers who are familiar with microservices cannot guarantee acceptable accuracy and efficiency with plain service call graphs. For detecting common anti-patterns, e.g., cyclic dependency and the knot, automated detection with interactive service call graphs is highly needed to facilitate such a practice for microservice-based systems, regardless of the system size.

The remainder of this paper is organized as follows: Section 2 presents the background and related work, Sect. 3 describes our visualization approach and the adopted format for the service view intermediate representation, Sect. 4 introduces the case study, and presents the results. Furthermore, Sect. 5 discusses the implication of this work as well as the threats to validity. Finally, Sect. 6 concludes the article.

2   Background and Related Work

Quality assurance safeguards are needed to fill gaps in the decentralized system environment to avoid microservice evolution leading to architecture degradation. One common avenue to identifying poor design quality is to identify anti-patterns [11]. A broad set of literature reviews are referenced in a recent tertiary study [5] specific to microservice anti-patterns. One option is to involve a human in the loop approach applied to an abstraction of the system to offset the effort needed to analyze the system as a whole.

Pattern detection has been approached using static [6,10,13] and dynamic analysis [1]. Typically, these approaches build a system intermediate representation and use it for this task when aiming for automated detection. While these publications looked into detection using static or dynamic analysis, the resulting prototype tools can only detect a small number of anti-patterns bound to technological constraints. Similarly, many dynamic analysis tools perform tracing and monitoring at the same time to extract a service call graph from the traces. While manual code review is impractical, one approach is to provide a visual representation of a simplified system model, such as the service call graph, to introduce a human-in-the-loop approach.

Conventional monitoring and tracing tools such as the previously mentioned Jaeger or Kiali (https://kiali.io) typically synthesize the service and operation views that consider the service call graphs, operation topology, and related concerns of the system. This view illustrates the whole system's interconnection and is especially common for microservices.

A possible solution to involving a human in the loop for analyzing anti-patterns in microservice systems is the creation of a visual service graph. The service view can be, to some extent, determined by dynamic analysis and static analysis. For instance, tools like OpenAPI (https://www.openapis.org) provide information about the endpoints and their signatures available per microservices. This can be combined with dynamic trace analysis to construct service call graphs. Similarly, a less precise but still feasible approach to determine this view is by analyzing source code and detecting inter-service remote method calls, and extracting their signatures which can be matched to microservice endpoints detected by Swagger. Dynamic analysis can add detail about the actual endpoint usage, and static analysis can add underlying details for endpoints such as access right resections (i.e., Java JSR-375), connected data entities (implied from data-flow analysis), etc. In summary, both static and dynamic analysis can uncover a service view.

The system service view is well suited to visualize anti-patterns. Literature mentions multiple options [4,16]. The established trajectory involves modeling languages such as UML, SysML, or TOGAF. However, none of these directions are interactive, which limits the ability to view a system from multiple angles or to isolate problems extending from particular services. We must assume that systems may involve hundreds of microservices, which can quickly become quite difficult for a human to understand from a single view.

3   Anti-pattern Visualization Framework for Microservices

A comprehensive approach to visualize the system's service view would start with static or dynamic system analysis. An *intermediate representation* of the system would then be constructed to encompass its structure. This could even be augmented later by extracting the necessary information from system artifacts.

To bridge the various service view extraction approaches, we emphasize the service view intermediate representation that works as a mediator data-interchange format. The service view intermediate representation could be produced by any means as illustrated in Fig. 1.

Information extraction can be approached in various ways to produce an intermediate representation that is fed as input to the visualization framework. This way, any third-party tool can use the visualization framework and benefit from its properties. In pursuit of this, our case study illustrates a combination with a third-party static analysis extraction approach.

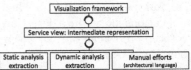

Fig. 1. Sketch or our framework in the context of extraction approaches

The system intermediate representation opens the framework to third-party integration providing our service view. It can accept manual model specification made for the purpose of anti-pattern demonstration, which is one anticipated long-term goal. An appropriate intermediate representation would use well-accepted and easy-to-transform data-interchange formats, such as JSON, YAML, or XML. It could be seen as a customized interface description language translated into the context of microservice architecture. We designed a custom JSON intermediate representation, which has been extended to a service view YAML language by a related work [7].

The intermediate representation recognizes deployable nodes by names, shapes, types (in most cases microservices, but possibly databases or message queues, etc.), endpoints, and oriented inter-service calls. The JSON format makes it easily extensible if necessary, and we imagine a service that could input varying models of a system to construct this representation. An example code snippet in Listing 1.1 describes how a microservice and its external connections are defined in the intermediate representation.

```
"nodes": [{
  "nodeName": "ts-preserve-other-service",
  "nodeType": "service",
  "dependencies": [],
  "targets": [{
    "nodeName": "ts-food-service",
    "requests": [{
      "type": "POST",
      "argument": "[@RequestBody FoodOrder addFoodOrder,
                    @RequestHeader HttpHeaders headers]",
      "msReturn": "org.springframework.http.HttpEntity",
      "endpointFunction": "foodsearch.controller
                            .FoodController.createFoodOrder",
      "path": "/FoodController/createFoodOrder"
    }],
  }]
}]
```

Listing 1.1. Intermediate representation example

The language preserves details of individual nodes (requests, type) while relating services together via a dependency graph. These details are necessary as several anti-patterns, such as low cohesive operations, require more detail than the type of service and its connections.

With intermediate representation as the input, the visualization of the service view aims to fit large systems with multiple modules and provide interactivity. When we analyzed existing visualizations, they used node-edge models in a flat view. Some visualizations support dragging the nodes around, but

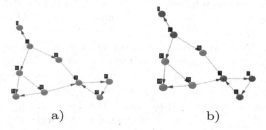

a)   b)

Fig. 2. a) Sample of a 9-service system. b) Illustrated highlight of cyclic dependency in the 9-service system.

they lack interactivity, which would enable quick node identification, highlighting searches, etc. There can we multiple levels of abstraction in the interactivity to navigate to necessary details hierarchically. We adopted an approach that has been used for semantic web representation with a node-edge model rendered in a 3D space due to the wide familiarity, availability, and maturity of the web medium.

The interactivity we considered had as the main perspective the ease of information identification. Thus we considered different node shapes and colors for different module types and problem highlights. We enabled interactivity by rearranging graph nodes or the view perspective (pan, zoom, and rotation). We also enabled the search feature to provide the ability to detail information about selected modules. Fig. 2-a illustrates our visualization of a simple system.

Next, we approached the identification of anti-patterns that can be illustrated from the service view perspective. For instance, we started with service coupling by integrating a slider for the threshold level and used color codes for the nodes in the model. The underlying graph traversal can easily identify particular metrics or make the operator aware of specific locations in the graph to draw his attention. This can be easily extended to other relevant metrics like structural coupling for microservices [9].

Anti-pattern detection involves specific rules to be checked for each pattern on the service view [2]. We did not anticipate a comprehensive approach and thus only illustrated a selected set of anti-patterns that could be extended by the community given dozens of microservice patterns and the open-source release of our frontend framework. The specific patterns we include are cyclic dependencies, bottleneck services, nano services, the knot, and duplicated services.

Cyclic dependencies exist when the external connections of one service allow it to reach other services and return back to itself. For example, if A targets B, B targets C, and C targets A, then A, B and C are all a part of a cyclic dependency. To detect this anti-pattern, we utilize the strongly connected components graph algorithm.

Bottleneck services are services that are being accessed by too many other services, thus slowing response time. The number of accesses that creates a slower response time is arbitrary and changes depending on the system. Some may find a high number of accesses to be acceptable in certain cases. This arbitrary number led us to implement a slider such that an individual using our visualizer can determine the threshold they deem to be too many. A node in the visualizer that depends on, meaning it is targeted by another node, more than the given threshold is determined to be a bottleneck service.

Nanoservices were also selected because of their similarity to bottleneck services. A nanoservice exists when a service does not provide a lot of utility but takes up resources because of its communication and maintenance. These are chosen to be detected by summing up the number of targets a node in the visualizer has and utilizing the user-defined threshold. We choose this method because a nanoservice typically sends out information to many different services since it is fulfilling functionality that other services do not implement.

The knot is another anti-pattern that deals with the coupling of services. If multiple services have high coupling with each other, then these services run more slowly and are likely performing tasks that could be consolidated. Since the knot is also concerned with nodes having too many relations, it also utilizes a user-defined threshold for detection. If the number of external relations a node has, plus the number of unique external relations its neighbors have, is greater than the threshold, then we say that node is a part of a knot.

Duplicated services implement similar functionality. These services should be combined into a larger service that takes the relations of the two existing services. We detect these duplicated services by finding services that have similar names. We choose this method because the intermediate representation of a node only defines the relations and the type of node.

Because we are visualizing microservices from a service view, we specifically want to focus on the issues resulting from inter-service communication between endpoints and service dependencies. For the initial prototype of the tool, we omitted visualization of microservice anti-patterns that arose due to coding practices during development and those that arose within individual microservices.

There are potential issues that arise with the defined detection techniques. First, services that have many external connections are not guaranteed to be detrimental to a system. We chose to detect the knot, bottleneck service, and nanoservice in this way because it is the most general way to detect these patterns. It is true that a node that is detected by our algorithm may be determined to not be an anti-pattern; however, a node that is a part of one of these anti-patterns is guaranteed to be found. Although our algorithm may return false positives, it will capture services that are creating an anti-pattern.

Second, two services are not guaranteed to be duplicated because they have similar names. Although the previous is true, we expect those utilizing this tool to understand a system and have some base knowledge of the said system. Thus they are able to make necessary judgment calls. For certain anti-patterns, such as duplicated service, where the detection is reliant on low-level information that is not described by an intermediate representation, we must find a detection technique that utilizes higher-level information. We chose services with similar names here because if two services are named properly and deliver similar functionality, they should have similar names.

Lastly, the anti-patterns that rely on user-defined thresholds should instead be detected in a more precise way. For example, the knot exists because multiple services that have low cohesion are coupled together. The knot does not exist because a node and its neighbors have more connections than a threshold. We chose to detect anti-patterns of this type with the number of external connections because low cohesion is different for every system. There is not a way, with an intermediate representation, to grasp the complexity of low cohesion. Therefore we give the user freedom to define the threshold as they see fit, again assuming that they have an understanding of the system.

4   Case Study Using Train Ticket System

We demonstrate a comprehensive perspective usage of our framework, in particular, the use of a microservice benchmark Train-Ticket [15] is established in the microservice community. This benchmark system is open-sourced with easy-to-understand business logic and explicitly established microservice architecture. It has also been commonly adopted by other researchers for microservice-related studies. For instance, Walker et al. [13] developed a tool that consumes the codebase, parses the code, and converts it into a component dependency graph intermediate representation. We have applied the same tools to extract the system's intermediate representation and performed a simple conversion to match the service view representation as an input to our visualization framework. The resulting interactive visualization is sketched in figures we share via https://zenodo.org/record/7671310. The system model can be rotated and zoomed (i.e., Zenodo Fig. 3) by the mouse/trackpad when seeking a detail; service details and connections are high-

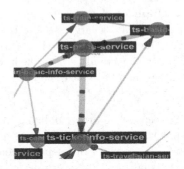

Fig. 3. Highlighted service connection detail

lighted when selected by the mouse cursor. It quickly becomes challenging to understand microservice systems even when reduced to an abstracted model parsed from the codebase. There is a large number of microservices and interconnections. When we consider an anti-pattern definition, it might be evident in small systems but becomes rather complicated to find manually in complex systems. When we aim to detect a metric setting for coupling, highlighting specific microservices drags the attention to the impacted interconnections, as illustrated in Fig. 3. When we consider anti-patterns, it is not obvious to detect them in the large system graph representation. An appropriate visualization, however, can help to narrow the user's attention using the illustrated service view. For instance, Fig. 2 b) illustrates a highlight of cyclic dependency on a fabricated service graph. Compared to the original sample Fig. 2-a, the cyclic dependencies are explicitly identifiable. While it is obvious that not all anti-patterns can be detected using the service view or the underlying system's intermediate representation, they can still be illustrated this way to aid users.

4.1　User Study

Intuitively, we expect an automated detection approach to be more efficient than manual effort to identify anti-patterns in a system.

Table 1. Participant familiarity with microservices

Familiarity with microservices	Number of participants	Cluster	Number of participants
Little to No Familiarity - 1	10		
2	2		
3	8	Mid-low	20
4	7		
High Familiarity - 5	1	High	8

The study results should demonstrate the potential impact of automated anti-pattern detection tools on the efficiency of tasks that are performed manually. It should also provide detail on the time needed to complete detection and recommend features that could help the user. Specifically, the study aims to tackle the following research questions.

RQ1. *How can anti-pattern visualization improve the precision of cyclic dependency detection using the service call graph?*

Cyclic dependencies can be detected manually by developers using a service call graph provided by tools such as Jaeger or Istio. In this RQ, we want to understand if developers are able to properly detect cyclic dependencies and how long it takes to detect them.

In order to evaluate the results of this RQ, we defined the following metrics:

- Number of correctly identified cyclic dependencies in a small system (True Positive)
- Number of correctly identified cyclic dependencies in a medium-sized system (True Positive)
- Number of wrongly identified cyclic dependencies in a small system (False Positive)
- Number of wrongly identified cyclic dependencies in a medium-sized system (False Positive)
- Time needed to identify the cyclic dependencies in a small system
- Time needed to identify the cyclic dependencies in a medium-sized system

RQ2. *How can anti-pattern visualization improve the precision of knot detection using the service call graph?*

Similar to cyclic dependencies, knots can be detected manually by developers starting from the service call graph. Following the same path, we want to understand if developers can properly detect the Knot and how long it takes to detect them.

In order to evaluate the results of this RQ, we defined the following metrics:

- Number of correctly identified instances of the knot in a small system (True Positive)
- Number of correctly identified instances of the knot in a medium-sized system (True Positive)
- Number of wrongly identified instances of the knot in a small system (False Positive)
- Number of wrongly identified instances of the knot in a medium-sized system (False Positive)
- Time needed to identify the instances of the knot in a small system
- Time needed to identify the instances of the knot in a medium-sized system

Case and Subjects Selection. In order to answer the RQs, we selected as cases four different microservice-based systems; two for cyclic dependencies and two for the knot. Both anti-patterns have small and medium-sized systems. Both small systems are artificially composed, while the medium systems are a modified subset of the Train-Ticket microservice benchmark system. For cyclic dependencies, the small system has 9 services and contains 2 cycles, while the medium system has 35 services and 3 cycles. The small system for the knot has 11 total services, 4 of them being included in the knot. The medium system for the knot contains 19 services, 5 of them being included in the knot. The service view of the 11-Service system is depicted in Figure 2-b while the view of the modified Train-Ticket with 35 services is available at https://zenodo.org/record/7671310.

The subjects involved in this study were junior developers with various levels of microservice domain knowledge. Each participant reported their programming experience and microservice domain knowledge. No personal information was collected or exposed during the process of the study.

Data Collection Procedure. To collect the data from the participants, we design the case study as follows. We recruited participants of various educational background from the authors' institutes. To classify the participants by their domain knowledge level, we asked the participants to identify their familiarity with microservices. Table 1 shows their familiarity on a scale of 1–5 (Little to No Familiarity – High Familiarity).

Firstly, we provide the overall introduction to the core concept of microservice architecture and the types of anti-patterns. We also introduce our visualizer with a demo for the participants to try out.

Entering the main tasks for each target anti-pattern, we introduce the concept as well as its detection method to the participants. In addition, illustrated examples with an 8-service system is provided before the task starts. The aim of the introduction is to guarantee each participant understands the detection mechanism correctly.

At the start, each participant is asked to report the anti-patterns in the small systems and the time they needed. The participants start and cease the timing themselves, which avoids any miscounting due to unawareness of the timing. After submitting their answers, the participants are given the correct answers and asked to check how many anti-patterns they correctly identified and their reflected insights. After completing the task on the small system, the participants work on the same task with the medium-sized modified Train-Ticket System with answers reported and reflected in the same fashion.

The participants are also asked to evaluate their levels of understanding of the anti-pattern after completing their detection on the small system. At the end of the study, the participants shall also reflect on their willingness to adopt a highlighted visualizer and provide potential comments and questions.

Data Analysis Procedure. Before the analysis, we preprocessed data separating the answers for the small and the medium systems analyzed. This allowed us to better compare all the answers with the different system sizes separately.

We did not convert ordinal data into numerical equivalents to avoid misleading results and to better identify the distribution of the answers.

For both RQs, we calculated the following metrics:

1. Familiarity with Microservices (Likert scale from 1 to 5; with 1 being very little familiarity and 5 being high familiarity)
2. The percentage of correctly identified cases (True Positive)
3. Number of wrongly identified cases in a small system (False Positive)
4. The average time taken to detect the anti-pattern
5. The standard deviation of the time taken to detect the anti-pattern

The analysis is performed for all the answers together, and then clustering answers by familiarity with Microservices.

4.2   Results

We gathered 28 responses in total, from two groups of participants: 8 individuals who are familiar with Microservices (either familiar or highly familiar), and 20 participants with medium to low familiarity. The median response among all participants for their familiarity with microservices was 3, indicating a medium level of familiarity. We have made the data available at https://zenodo.org/record/7676003.

Note that 10 min was the upper bound for the Train-Ticket systems. Three cycles were correct in the 35-service system, see Fig. 4, and five services were correct in the 19-service system. Finally, we analyzed the feedback for outstanding comments.

In order to understand if the visualization of anti-patterns improves the precision of the detection of cyclic dependency (**RQ1**), we analyzed the answers to the first task. The results in Table 2 show that manual detection of anti-patterns is not precise given a service call graph.

As shown in the results, participants can only correctly identify, on average, 66% to 70% of the cyclic dependencies. Even with the small system, only 32% of the participants managed to identify all cyclic dependencies correctly (46% on the medium-sized system).

Considering the system size, on the 9-service system, very few participants reported a false positive (4%), and the analysis of the system was performed in 1 min 23 s on average (± 46 s); For the medium-sized system, there is not a significant change in true positives. The number of false positives, however, increases in the medium system (30%). The time taken to detect cyclic dependencies in the medium-sized system also increases (5 min 13 s ± 2 min 2 s).

We also considered the perspective based on participant familiarity with microservices as illustrated in Table 3 and Table 4. For the small system, the participants that were more familiar performed better with respect to accurate pattern detection (75% compared to 63%). The same comparative outcome was shown regarding the medium-sized system (79% to 67%). However, they spent less time on achieving the results (1 min 1 s compared to 1 min 31 s on average) with

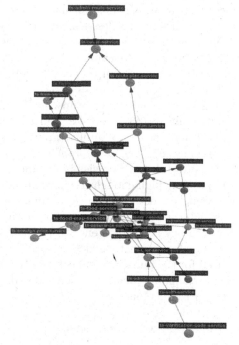

Fig. 4. Automated cyclic-dependency detection applied on the Train-Ticket system

the small system but more time with the medium system (5 min 3 s to 5 min 39 s on average). Thus, it could be understood that familiarity with microservices does not reduce the time needed to perform cyclic dependency detection. We similarly found that the high familiarity group performed better than the lower familiarity group (50% vs. 25% for the small system and 63% compared to 40% for the medium system).

While considering the knot (**RQ2**), participants correctly identified on average 72% of the patterns in the small (11 services) system and 53% of the patterns in the medium train-ticket system (19 services). For the 11-service system, 22% of participants reported false positives, averaging 0.35 services wrongly identified per participant. The 19-service system had a greater number of false positives, averaging 0.61 wrongly identified services as knots (39% of participants). Additionally, detecting knots in the complex system required a greater amount of time when compared to the simple system. Participants spent an average of 2 min 27 s ± 2 min 2 s on the 11-service system, compared to an average of 4 min 23 s ± 2 min 50 s for the 19-service system.

Considering the difference in participants' familiarity with microservices, the more familiar cluster surprisingly performed worse for the small knotted system compared to the less familiar cluster (68% to 74%). On the contrary, with the medium system, their performances are the opposite (60% to 50%). The more familiar cluster spent less time on achieving the results (2 min 24 s compared to 2 min 28 s on average) with the small system but more time with the medium system (4 min 43 s to 4 min 14 s on average). Participants who correctly identified all anti-patterns encompassed 44% for the 11-service system; meanwhile, only 17% of participants were able to correctly identify all services for the 19-service system. Between the two groups, low familiarity and high familiarity participants fared similarly (43% and 44%) in correctly identifying all knots on the small system, but the more familiar group was better at identifying all knots on the medium system (29% compared to 13%).

To prove that the precision of manual detection of anti-patterns with a service call graph is high enough to eliminate the need for an automatic detection tool, we would expect there to be minimal error in detection. As the level of error was high considering the mean familiarity with microservices, we determine that the accuracy is not high enough to prove that service call graphs provide enough precision for anti-pattern detection. We gather from this information that there is a need for the automatic detection and visualization of anti-patterns. Regarding the participants' open comments on the framework, a set of pros and cons are summarized as follows.

The framework pros include:

+ Highlighting the service call with dynamic arrows
+ Zooming, clustering, rotating and moving around
+ Straightforward and ease-of-use

The framework cons include:

- Hard to read when system size increases
- Cannot delete, filter, or drag the nodes

5   Discussion

The results of this work show the effectiveness of using an interactive service call graph for the manual detection of cyclic dependencies and the knot in microservice-based systems.

Table 2. Results of the case study

RQs	System Size	Correct Answers	Correctly Identified Cases		Wrongly Identified Cases	Time	
			Mean#	Mean%	Mean#	AVG	σ SD
RQ1 Cyclic Dependency	9-microservice	2	1.32	66%	0.04	00:01:23	00:00:46
	35-microservice	3	2.11	70%	0.29	00:05:13	00:02:02
RQ2 Knot	11-microservice	4	2.87	72%	0.35	00:02:27	00:01:22
	19-microservice	5	2.70	53%	0.61	00:04:23	00:02:50

Regarding the detection of cyclic dependency (RQ1), our case study shows that manual detection is not accurate enough. It is hard for the developers, regardless of their familiarity of microservice architecture, to detect all the cyclic dependencies. Meanwhile, the developers with better familiarity spent more time on completing the detection tasks than the ones with less familiarity. Regarding the detection of the knot (RQ2), the results are similar in terms of the low accuracy in manual detection and more time needed with increased system size. Surprisingly, the developers with less microservice familiarity performed better than the ones more familiar with the small system. Both, however, fall short in the accuracy of the medium system. Developers with better familiarity spent more time completing the detection tasks with the medium system. It can imply that familiarity with microservices does not reduce time needed to manually detect both cyclic dependencies and the knot.

The implications of this study are as follows.

1. This study shows it is necessary to automate anti-pattern detection for microservice-based systems, as the average accuracy of detecting anti-patterns is not satisfactory.
2. It also shows the developers' familiarity with microservice in general, does not reduce the time needed for anti-pattern detection. The performance of the ones with higher familiarity still requires improvement.
3. Furthermore, interactivity is appreciated by the majority of the developers who require more interactivity with the service call graphs with better readability for large-size systems.
4. In addition, the prototype framework could serve operators when combined with analytical tools to identify quality problems in their systems.

Due to the low accuracy of manual anti-pattern detection, the need for automatic detection feature is needed for service call graphs. Furthermore, based on developer feedback, more interactivity to the service graph can improve the usability of the framework. Enhanced readability for medium and large-sized systems is also required.

5.1   Threats to Validity

This section addresses the threats to the validity of our research. We consider Wohlin's taxonomy [14] for this.

Table 3. Results of the case study for 8 microservice highly-familiar participants (> 3)

RQs	System Size	Correct Answers	Correctly Identified Cases		Wrongly Identified Cases	Time	
			Mean#	Mean%	Mean#	AVG	σ SD
RQ1 Cyclic Dependency	9-microservice	2	1.50	75%	0.00	00:01:01	00:00:31
	35-microservice	3	2.38	79%	0.00	00:05:39	00:01:33
RQ2 Knot	11-microservice	4	2.71	68%	0.14	00:02:24	00:00:51
	19-microservice	5	3.00	60%	0.14	00:04:43	00:02:19

Table 4. Results of the case study for 20 participants with medium or low familiarity with microservices (≤ 3)

RQs	System Size	Correct Answers	Correctly Identified Cases		Wrongly Identified Cases	Time	
			Mean#	Mean%	Mean#	AVG	σ SD
RQ1 Cyclic Dependency	9-microservice	2	1.25	63%	0.05	00:01:31	00:00:48
	35-microservice	3	2.00	67%	0.40	00:05:03	00:02:13
RQ2 Knot	11-microservice	4	2.94	74%	0.44	00:02:28	00:01:33
	19-microservice	5	2.50	50%	0.81	00:04:14	00:03:06

Table 5. Participants who correctly identified all system anti-patterns by familiarity

Group	Cyclic Dependency				Knot			
	9-Microservice		35-Microservice		11-Microservice		19-Microservice	
	Participants	%	Participants	%	Participants	%	Participants	%
High Familiarity	4	50%	5	63%	3	43%	2	29%
Low Familiarity	5	25%	8	40%	7	44%	2	13%
All	9	32%	13	46%	10	44%	4	17%

Construct Validity: We have implemented a custom service view visualization, which is based on similar models used by established industry tools. With regards to system size, we chose to develop the tool in 3D space. It is a node-edge graph with oriented edges, which most computer scientists are familiar with. One construct threat to validity was that we illustrated a cycle between two nodes by a bi-directional arrow, rather than two arrows. This situation is illustrated in Fig. 5. This was mentioned in feedback from participants as a source of errors, however, they were able to learn for the medium-sized system. Similarly, node labels can overlap with arrows and could impact detection.

Regarding the user study, we followed conventional practices for such studies. It involved survey forms, timed tasks, and user opinions.

Fig. 5. Used bi-directional arrow

One concern about the conduction of our case study is that the manual detection was performed using our system without the highlighting feature. We asked the participants to manually perform the detection on our tool so that they would only need to learn one tool. It is reasonable to suggest that participants could have been more accurate with a different tool. We negate this concern because regardless of the system that generates a service call graph, the information presented would have been reasonably the same.

Internal Validity: Regarding the user study, all participants performed the entire study in the same order. We tracked their familiarity with microservices in different form, which we later connected to the study records. We considered that the small-sized experiments could serve as training for the medium-sized study. The primary goal, however, was to consider the accuracy and time needed to perform the detection. To avoid user memorization when performing all case studies, we used different system versions. It is important to mention that this experiment was performed remotely. The study results, however, depend on self-reporting. Similarly, at the beginning of each session, we asked participants if they understood the anti–pattern definition to understand how it may have impacted their use of the visualizer.

All participants were instructed to utilize a computer rather than a mobile device in order to better control variety in medium that the visualizer was utilized in. We did not restrict users to using a mouse or trackpad and did not collect data on the user's device properties.

External Validity: Regarding the user study, we have used a third-party system benchmark and its versions recognized by the community to avoid bias. We reached a diverse spectrum of practitioners with different levels of familiarity with microservices. When reasoning about experiment results, we divided the perspectives of the user familiarity with microservices.

Conclusion Validity: To strengthen research and claims, we performed a user study with unbiased participants with different expertise levels. The case study generated insights for designing similar visualization tools, limitations, and open challenges to be addressed when using the 3D approach in practice.

6 Conclusions

The service call graph is an established perspective used by many microservice monitoring tools and also facilitates anti-pattern detection. We developed a framework that uses a system intermediate representation for the service call graph. We illustrate in this framework the capability to visualize microservice systems in an interactive manner to allow the detection of anti-patterns. The open-sourced prototype is meant to encourage the community in using it to illustrate microservice anti-patterns to train an expert workforce, or to use it for detection of anti-patterns from arbitrary sources. The results shows that despite an interactive visualization of anti-patterns, manual detection is not precise enough to remove a need for an automatic detection tool. Regardless of background and prior experience with microservices and anti-patterns, participants took a significant amount of time to identify anti-patterns and often struggled to identify all occurrences in a system. Given that these services were only small and medium-sized, it would soon become impossible for a human to manually identify anti-patterns in an industry-scale system of hundreds of nodes.

As for future works, we shall further explore customized architectural languages for our intermediate representation, provenance tracking for operator navigation history, and a shared view perspective of multi-observers. Additionally, we will endeavor to improve the prototype against large systems and augment its features to take into account the feedback discovered during the case study. Moving towards potential industrialization, we also consider the future integration with OpenTelemetry analytical tools, usability enhancement, and continuous evolution with CI/CD pipeline connection.

Acknowledgments. This material is based upon work supported by the National Science Foundation under Grant No. 1854049 and a grant from Red Hat Research, https://research.redhat.com.

References

1. Bakhtin, A., Al Maruf, A., Cerny, T., Taibi, D.: Survey on tools and techniques detecting microservice API patterns. In: IEEE International Conference on Services Computing (SCC) (2022)
2. Bogner, J., Boceck, T., Popp, M., Tschechlov, D., Wagner, S., Zimmermann, A.: Towards a collaborative repository for the documentation of service-based antipatterns and bad smells. In: 2019 IEEE International Conference on Software Architecture Companion (ICSA-C), pp. 95–101. IEEE (2019)
3. Bogner, J., Fritzsch, J., Wagner, S., Zimmermann, A.: Industry practices and challenges for the evolvability assurance of microservices. Empir. Softw. Eng. **26**(5), 104 (2021). https://doi.org/10.1007/s10664-021-09999-9
4. Cerny, T., Abdelfattah, A., Bushong, V., Maruf, A.A., Taibi, D.: Microservice architecture reconstruction and visualization techniques: a review. In: 2022 IEEE Symposium on Service-Oriented System Engineering (SOSE) (2022)
5. Cerny, T., Al Maruf, A., Janes, A., Taibi, D.: Microservice anti-patterns and bad smells. how to classify, and how to detect them. a tertiary study (2023)
6. Fontana, F.A., Pigazzini, I., Roveda, R., Tamburri, D., Zanoni, M., Di Nitto, E.: Arcan: a tool for architectural smells detection. In: 2017 IEEE International Conference on Software Architecture Workshops (ICSAW), pp. 282–285 (2017). https://doi.org/10.1109/ICSAW.2017.16

7. Lelovic, L., Mathews, M., Abdelfattah, A.S., Cerny, T.: Microservices architecture language for describing service view. In: 13th International Conference on Cloud Computing and Services Science (CLOSER) (2023)

8. Palma, F., Mohay, N.: A study on the taxonomy of service antipatterns. In: 2015 IEEE 2nd International Workshop on Patterns Promotion and Anti-patterns Prevention (PPAP), pp. 5–8. IEEE (2015)

9. Panichella, S., Rahman, M.I., Taibi, D.: Structural coupling for microservices. arXiv preprint arXiv:2103.04674 (2021)

10. Pigazzini, I., Fontana, F.A., Lenarduzzi, V., Taibi, D.: Towards microservice smells detection. In: Proceedings of the 3rd International Conference on Technical Debt, TechDebt '20, pp. 92–97 (2020)

11. Taibi, D., Lenarduzzi, V.: On the definition of microservice bad smells. IEEE Softw. **35**(3), 56–62 (2018). https://doi.org/10.1109/MS.2018.2141031

12. Taibi, D., Lenarduzzi, V., Pahl, C.: Microservices anti-patterns: a taxonomy. In: Bucchiarone, A., et al. (eds.) Microservices: Science and Engineering, pp. 111–128. Springer, Cham (2020). https://doi.org/10.1007/978-3-030-31646-4_5

13. Walker, A., Das, D., Cerny, T.: Automated code-smell detection in microservices through static analysis: a case study. Appl. Sci. **10**(21) (2020). https://doi.org/10.3390/app10217800

14. Wohlin, C., Runeson, P., Höst, M., Ohlsson, M.C., Regnell, B., Wesslén, A.: Experimentation in Software Engineering. Springer, Berlin, Heidelberg (2012). https://doi.org/10.1007/978-3-642-29044-2

15. Zhou, X., et al.: Benchmarking microservice systems for software engineering research. In: Proceedings of the 40th International Conference on Software Engineering: Companion Proceeedings (2018)

16. Zhou, Z., Zhi, Q., Morisaki, S., Yamamoto, S.: A systematic literature review on enterprise architecture visualization methodologies. IEEE Access **8**, 96404–96427 (2020). https://doi.org/10.1109/ACCESS.2020.2995850

An Accurate and Preservative Quenching Data Stream Simulation Method

Eduardo Servin Torres$^{(\boxtimes)}$ and Qin Sheng

Department of Mathematics and Center for Astrophysics, Space Physics and
Engineering Research, Baylor University, Waco, TX 76798-7328, USA
{Eduardo_Servin1,Qin_Sheng}@baylor.edu

Abstract. Quenching has been an extremely important natural phenomenon observed in many biomedical and multiphysical procedures,
such as a rapid cancer cell progression or internal combustion process.
The latter has been playing a crucial rule in optimizations of modern solid
fuel rocket engine designs. Mathematically, quenching means the blow-
up of temporal derivatives of the solution function q while the function
itself remains to be bounded throughout the underlying procedure. This
paper studies a semi-adaptive numerical method for simulating solutions
of a singular partial differential equation that models a significant number of quenching data streams. Numerical convergence will be investigated as well as verifying that features of the solution is preserved in the
approximation. Orders of the convergence will also be validated through
experimental procedures. Milne's device will be used. Highly accurate
data models will be presented to illustrate theoretical predictions.

1 Introduction

A key behavior observed during tumor progress, wound healing, and cancer inva-
sion is that of rapid collective and coordinated cellular motion. Hence, understanding the different aspects of such coordinated migration is fundamental for
describing and treating cancer and other pathological defects [1,2]. To reduce
the number of invasive surgical procedures on patients, accurate tumor models
and simulations have become crucial in the study. One of such effective models
is built the nonlinear quenching partial differential equation which characterizes
sudden growths of cancer cells once certain environmental criteria are reached.
Similar modeling equations are frequently used in the energy industry for internal combustion machine designs [3–5].

For the sake of simplicity in formulations, we focus at an one-dimensional
quenching model problem in this paper. In the circumstance, the quenching
dynamics can be characterized through following modeling problem [3,6–8]:

This research is supported in part by the National Science Foundation (grant No.
DMS-2318032; USA) and Simons Foundation (grant No. 1001466; USA).

$$\sigma(s)q_t = \frac{1}{a^2}q_{ss} + \phi(q), \quad 0 < s < 1, \; t_0 < t \leq T, \tag{1.1}$$

$$q(0,t) = q(1,t) = 0, \quad t > t_0, \tag{1.2}$$

$$q(s,t_0) = q_0(s), \quad 0 \leq s \leq 1, \tag{1.3}$$

where $a > 0$ is the physical size of a tumor contaminated region, or a linear combustion chamber, q is the cell population index, $\phi(q) \to +\infty$ as $q \to b^-$, b is a trigging threshold of the population, and $T < +\infty$ is sufficiently large. We adopt the following degenerate and reaction functions,

$$\sigma(s) = \alpha s^\theta (1-s)^{1-\theta}, \; \phi(q) = (b-q)^{-p}, \quad \alpha > 0, \; 0 \leq \theta \leq 1, \; p > 0. \tag{1.4}$$

Note that $\sigma(s) = 0$ indicates possible hidden defects within combustion chamber walls, and locations of such defects can be stochastically distributed in the spacial domain. Size of such a location set is often extremely small otherwise they can be detected in earlier stage of the manufacturing process [4,6,9]. On the other hand, the nonlinear source function $\phi(q)$ must be monotonically increasing with $\phi(0) = \phi_0 > 0$ and $\lim_{q \to b^-} \phi(q) = \infty$. Our functions in (1.4) are particularly chosen to reflect aforementioned features in a relatively simple manner, and to achieve quick and successful data stream analysis in mathematics.

It has been shown that there exists a *critical value* $a^* > 0$ such that if a in (1.1) is greater than a^* then the maximal value of solution of (1.1)–(1.3) reaches its ceiling b in finite time $T_a = T(a)$. This indicates that

$$\lim_{t \to T_a^-} \max_{s \in [0,1]} q(s,t) = b \quad \text{and} \quad \lim_{t \to T_a^-} \sup_{s \in (0,1)} q_t(s,t) = +\infty.$$

Such a phenomenon is refereed to as *quenching,* and the corresponding q is a *quenching solution.* Further, q must increase monotonically as t increases at any fixed cell location $0 < s < 1$ [10–13].

In the study of numerical combustion, the quenching stream (1.1)–(1.3) is particularly used to model combustible systems utlizing solid or liquid fuels. The ignition process starts with appearance of a outside thermal source which results in an region heating up. If the conditions are appropriate then the region will have high temperatures with drastic increase in reaction rates, eventually resulting in an explosion. The process may be found in everyday applications like automobile engine and in a more interesting setting, rocket engines. In addition to the function ω used to show certain defects in side-wall of a combustor, air bubbles contained in the fuel and, more seriously, hidden cracks in engine structures can also be formulated approximately. The partial differential equations often provide lower cost evaluations of modern engine designs before any expensive physical tests and experiments. The mathematical model and data obtained also help optimize the improvement of engines to maximize the efficiency in fuel consumptions. Needless to say, this has been one of the top concerns in the energy industry.

Our investigation is organized as follows. In the next section, we propose a second order Crank-Nicolson scheme for solving (1.1)–(1.4) on uniform spatial mesh and adaptive temporal steps. Preservations of the solution geometry such as the cell population positivity and monotonicity are studied. A proof of the convergence of the numerical solution sequence is given. A remark is stated for the more general simulation applications. Section 3 focuses on multiple numerical experiments that illustrate our analysis. Comparisons are offered with typical numerical methods in the field. Experiments are conducted on the order of accuracy of our simulation method. It is found that the method remarkably retains second order accuracy in space and first order in time, except in the quenching, or cell bursting, area as the quenching time is approached. Finally, in Sect. 4, brief concluding remarks and discussions are given for future endeavors in biomedical simulations.

2   Conservative and Convergent Algorithm

Let $b = 1$, $N \in \mathbb{N}^+$, $N \gg 1$ and $h = 1/(N+1)$. Further, let $\bar{\mathcal{D}}_N = \{s_0, s_1, \ldots, s_{N+1}\} \subset \bar{\Omega}$, where $s_k = kh$, $k = 0, 1, \ldots, N+1$. Denote $q_k^{(i)}$ as an approximation of $q(s_k, t_i)$, $k = 0, 1, \ldots, N+1$, $i = 0, 1, \ldots$ Assume that $\mathcal{D}_N \subset \bar{\mathcal{D}}_N$ be the set of interior mesh points. We approximate the spacial derivative in (1.1) through second-order central difference

$$(q_{ss})_k^{(i)} = \frac{q_{k-1}^{(i)} - 2q_k^{(i)} + q_{k+1}^{(i)}}{h^2} + \mathcal{O}(h^2), \quad s_k \in \mathcal{D}_N.$$

Drop the truncation error. Utilizing a Crank-Nicolson method we obtain the following semi-adaptive nonlinear method from (1.1)–(1.3):

$$q^{(j+1)} = \left(I - \frac{\tau_j}{2}A\right)^{-1}\left(I + \frac{\tau_j}{2}A\right)\left[q^{(j)} + \frac{\tau_j}{2}\psi\left(q^{(j)}\right)\right] + \frac{\tau_j}{2}\psi\left(q^{(j+1)}\right), \quad j = 0, 1, \ldots, J, \quad (2.1)$$

$$q^{(0)} = q_0, \quad (2.2)$$

where $q^{(i)} = \left(q_1^{(i)}, q_2^{(i)}, \ldots, q_N^{(i)}\right)^\top$, $\psi = \left(\frac{\phi_1^{(i)}}{\sigma_1}, \frac{\phi_2^{(i)}}{\sigma_2}, \ldots, \frac{\phi_N^{(i)}}{\sigma_N}\right)^\top$, $A = BT \in \mathbb{R}^{N \times N}$,

$$B = \text{diag}\left[\frac{1}{\sigma_1}, \frac{1}{\sigma_2}, \ldots, \frac{1}{\sigma_N}\right], \quad T = \frac{1}{a^2 h^2}\begin{bmatrix} -2 & 1 & & & \\ 1 & -2 & 1 & & \\ & \cdots & \cdots & \cdots & \\ & & 1 & -2 & 1 \\ & & & 1 & -2 \end{bmatrix}, \quad k = 1, 2, \ldots, N,$$

and $q_k^{(\ell)}$ is an approximation of $q(t_\ell)$, $t_\ell = \sum_{k=0}^{\ell} \tau_k$, $\ell = 0, 1, 2, \ldots, j+1$, and variable temporal steps τ_j can be determined through a proper monitoring function, such as an arc-length function [6,14]. Needless to mention, an iterative procedure, or a linearization of the last term in (2.1), needs to be implemented for solving system (2.1), (2.2). Monotone upper-lower solution vector procedures may also be incorporated in such computations [11].

As discussed intensively in [12,15,16], solution positivity and monotonicity are among the most distinguished mathematical characteristics of singular problems such as (1.1)–(1.3), and thus (2.1), (2.2). These properties reflect proper natural behaviors of the cancer cell population growth or decay, and should be preserved throughout simulations. To the end of analysis, we let $\vee$ be one of the operations $<$, $\leq$, $>$, $\geq$. For α, $\beta \in \mathbb{R}^N$, we assume following notations:

1. $\alpha \vee \beta$ means $\alpha_k \vee \beta_k$, $k = 1, 2, \ldots, N$;
2. $c \vee \alpha$ means $c \vee \alpha_k$, $k = 1, 2, \ldots, N$, for any $c \in \mathbb{R}$.

If $\frac{\tau_j}{h^2} \leq \frac{2a^2}{\sigma_{\max}}$, $j \in \mathbb{N}$. Then matrices $I - \frac{\tau_j}{2}A$, $I + \frac{\tau_j}{2}A$ are nonsingular. Furthermore, $I + \frac{\tau_j}{2}A$ is nonnegative, $I - \frac{\tau_j}{2}A$ is monotone and inverse-positive. Under the same constraint, If there exits $\ell > 0$ such that

$$\frac{\tau_j \psi_q(\xi_k^{(j)})}{2} \leq 1, \ k = 1, 2, \ldots, N; \ Aq^{(j)} + \psi\left(q^{(j)}\right) \geq 0, \ j = 0, 1, \ldots, \ell,$$

then the solution sequence, $q^{(0)}$, $q^{(1)}, \ldots, q^{(\ell)}, \ldots$, generated by (2.1), (2.2) are monotonically increasing. To see the above, from (2.1) we may observe that

$$q^{(j+1)} - q^{(j)} = \left(I - \frac{\tau_j}{2}A\right)^{-1}\left(I + \frac{\tau_j}{2}A\right)\left[q^{(j)} + \frac{\tau_j}{2}\psi\left(q^{(j)}\right)\right]$$
$$+ \frac{\tau_j}{2}\psi\left(q^{(j+1)}\right) - q^{(j)} = \left(I - \frac{\tau_j}{2}A\right)^{-1} w^{(j)}, \qquad (2.3)$$

where

$$w^{(j)} = \tau_j Aq^{(j)} + \tau_j \psi\left(q^{(j)}\right) + \frac{\tau_j}{2}\left(I - \frac{\tau_j}{2}A\right)\psi_q\left(\xi^{(j)}\right)\left(q^{(j+1)} - q^{(j)}\right).$$

Substitute the above back into (2.3) to yield

$$q^{(j+1)} - q^{(j)} = \tau_j \left[I - \frac{\tau_j}{2}\psi_q\left(\xi^{(j)}\right)\right]^{-1}\left(I - \frac{\tau_j}{2}A\right)^{-1}\left[Aq^{(j)} + \psi\left(q^{(j)}\right)\right].$$

Therefore $\left[I - \frac{\tau_j}{2}\psi_q\left(\xi^{(j)}\right)\right]^{-1}$ is nonnegative. This ensures the anticipated monotonicity $q^{(j+1)} \geq q^{(j)}$.

Theorem A. *The semi-adaptive method (2.1), (2.2) is convergent.*

Proof. Assume that $Q_k^{(\ell)}$ be the exact solution of (1.1)–(1.3), then

$$Q^{(j+1)} = \left(I - \frac{\tau_j}{2}A\right)^{-1}\left(I + \frac{\tau_j}{2}A\right)\left[Q^{(j)} + \frac{\tau_j}{2}\psi\left(Q^{(j)}\right)\right] + \frac{\tau_j}{2}\psi\left(Q^{(j+1)}\right) + \mathcal{O}\left(\tau_j^2\right).$$

Subtracting (2.1) from the above and denote $\varepsilon^{(\ell)} = Q^{(j+1)} - q^{(j+1)}$, we find that

$$\varepsilon^{(j+1)} = \left(I - \frac{\tau_j}{2}A\right)^{-1}\left(I + \frac{\tau_j}{2}A\right)\left\{\varepsilon^{(j)} + \frac{\tau_j}{2}\left[\psi\left(Q^{(j)}\right) - \psi\left(q^{(j)}\right)\right]\right\}$$
$$+ \frac{\tau_j}{2}\left[\psi\left(Q^{(j+1)}\right) - \psi\left(q^{(j+1)}\right)\right] + \mathcal{O}\left(\tau_j^2\right). \tag{2.4}$$

Note that

$$\psi\left(Q^{(\ell)}\right) - \psi\left(q^{(\ell)}\right) = \psi_q\left(\xi^{(\ell)}\right)\varepsilon^{(\ell)},$$

where elements of $\xi^{(\ell)}$, $\xi_k^{(\ell)} \in \left(\min\left\{q_k^{(\ell)}, Q_k^{(\ell)}\right\}, \max\left\{q_k^{(\ell)}, Q_k^{(\ell)}\right\}\right)$, $k = 1, 2, \ldots, N$. Recall (2.1). From (2.4) we obtain immediately that

$$\varepsilon^{(j+1)} = \left(I - \frac{\tau_j}{2}E_1^{(j+1)}\right)^{-1}\left(I - \frac{\tau_j}{2}A\right)^{-1}\left(I + \frac{\tau_j}{2}A\right)\left(I + \frac{\tau_j}{2}E_0^{(j)}\right)\varepsilon^{(j)} + \mathcal{O}\left(\tau_j^2\right).$$

It follows readily that

$$\left\|\varepsilon^{(j+1)}\right\|_2 = \left\|\left(I - \frac{\tau_j}{2}E_1^{(j+1)}\right)^{-1}\left(I - \frac{\tau_j}{2}A\right)^{-1}\left(I + \frac{\tau_j}{2}A\right)\left(I + \frac{\tau_j}{2}E_0^{(j)}\right)\varepsilon^{(j)}\right\|_2 + C_0\tau_j^2,$$

where $C_0 > 0$ is a constant. Using norm properties,

$$\left\|\varepsilon^{(j+1)}\right\|_2 \le (1 + M\tau_j)\left\|\varepsilon^{(j)}\right\|_2 + C\tau_j^2,$$

where constants $M, C \in \mathbb{R}^+$. Use the above inequality recursively. We acquire that

$$\left\|\varepsilon^{(j+1)}\right\|_2 \le (1 + M\tau_j)(1 + M\tau_{j-1})\left\|\varepsilon^{(j-1)}\right\|_2 + (1 + M\tau_j)C\tau_{j-1}^2 + C\tau_j^2$$
$$= (1 + M\tau_j)\ldots(1 + M\tau_0)\left\|\varepsilon^{(0)}\right\|_2 + C\tau_0^2(1 + M\tau_j)\ldots(1 + M\tau_1) + \cdots$$
$$+ C\tau_{j-1}^2(1 + M\tau_j) + C\tau_j^2$$
$$= C\tau_0^2(1 + M\tau_j)\cdots(1 + M\tau_1) + \cdots + C\tau_{j-1}^2(1 + M\tau_j) + C\tau_j^2,$$

since $\varepsilon^{(0)} = 0$ due to the initial value used. We further observe that

$$C\tau_0^2(1 + M\tau_j)\cdots(1 + M\tau_1) + \cdots + C\tau_{j-1}^2(1 + M\tau_j) + C\tau_j^2$$
$$= C\tau_0^2(1 + M\tau_0^*)^j + C\tau_1^2(1 + M\tau_1^*)^{j-1} + \cdots + C\tau_{j-1}^2(1 + M\tau_{j-1}^*) + C\tau_j^2,$$

where τ_i^* for $i = 0, 1, \ldots, j-1$ are adaptive time steps used. Recursively, we obtain that

$$C\tau_0^2(1 + M\tau_0^*)^j + C\tau_1^2(1 + M\tau_1^*)^{j-1} + \cdots + C\tau_{j-1}^2(1 + M\tau_{j-1}^*) + C\tau_j^2$$
$$\le C\tau_0^2 e^{b_0 T} + C\tau_1^2 e^{b_1 T} + \cdots + C\tau_{j-1}^2 e^{b_{j-1} T} + C\tau_j^2,$$

where b_i for $i = 0, 1, \ldots, j - 1$ are positive constants such that $\tau_i^* = b_i \tau_{b_i}^*$ and $\tau_{b_i}^* = T/N$. Let us take the b_k that gives us the maximum of b_i in the above. Therefore,

$$C\tau_0^2 e^{b_0 T} + C\tau_1^2 e^{b_1 T} + \cdots + C\tau_{j-1}^2 e^{b_{j-1} T} + C\tau_j^2 \leq C e^{b_k T} (\tau_0^2 + \tau_1^2 + \cdots + \tau_{j-1}^2 + \tau_j^2).$$

Now, denote $C e^{b_k T} = \tilde{C}$ and set $\tau_m = \max_\ell \tau_\ell \ll 1$. From above investigations, we have

$$\left\| \varepsilon^{(j+1)} \right\|_2 \leq \tilde{C}\tau_m (\tau_0 + \tau_1 + \cdots + \tau_{j-1} + \tau_j) = \tilde{C}\tau_m T \to 0$$

as $h, \tau_m \to 0$. This ensures the expected convergence. ∎

Remark 2.1. As an extension of the theorem, we may also prove that for any given $\ell \in \mathbb{N}$ and beginning solution $q^{(\ell)} < 1$, if

$$\gamma_j = \frac{\tau_j}{h^2} \leq \frac{2a^2}{\sigma_{\max}}; \ \frac{\tau_j}{2} \psi_q(\xi_k^{(j)}) \leq 1, \ k = 1, 2, \ldots, N; \ Aq^{(j)} + \psi\left(q^{(j)}\right) \geq 0, \ j = \ell, \ell+1, \ell+2, \ldots,$$

then the vector solution sequence $q^{(\ell)}$, $q^{(\ell+1)}$, $q^{(\ell+2)}, \ldots$, generated by the semi-adaptive scheme (2.1), (2.2) increases monotonically until unity is exceeded by a component of the vector (that is, until quenching occurs) or converges to a steady solution of the problem (1.1)–(1.3). In the latter case, we do not have a quenching solution.

Remark 2.2. We note that the last set of inequalities used in Remark 2.1 has been ensured at least for the case $\ell = 0$ and $q^{(0)}$. It seems that the solution monotonicity requires more rigorous constraints than those for the numerical convergence. This additional numerical feature is definitely justified for ensuring expected quenching-blow up phenomena. However, the quenching data monotonicity requirement has also made applications of nonuniform spacial grids much more challenging.

Remark 2.3. The simulation method (2.1), (2.2) is numerical stable in the von Neumann sense [17]. Since the scheme is often solved via a suitable linearization of the nonlinear function ψ, the simulation method is also convergent due to a natural conclusion of the Lax equivalence theorem. However, in the circumstance, the order of convergence remains to be determined [11,18].

3 Order of Convergence and Simulation Results

The order of convergence $r > 0$ of a consistent numerical method is defined through the error estimate,

$$\left\| q^{(j)} - q(t_j) \right\|_p = \mathcal{O}(h^r), \quad h \to 0^+, \ j \in \{0, 1, \ldots, J\}, \tag{3.1}$$

in the p-norm ($p \geq 1$). Since (3.1) is in general difficult to use, r is often replaced practically by the order of the truncation error, or defect, of the underlying method [5,18,19].

Theorem B. *The order of convergence of the semi-adaptive method* (2.1), (2.2) *is quadratic in the maximum norm.*

Proof. Since our simulation method (2.1), (2.2) is proven to be convergent and numerical stable according to the Lax equivalence theorem, we know that $r > 0$. To estimate such an order via a defect function, we notice that

$$\sigma(s)q_t - \frac{1}{a^2}q_{ss} - \phi(q) = 0$$

due to (1.1). Further, let τ, h be the temporal and spacial discretization parameters, respectively. Based on the forward temporal difference and central spacial difference, we acquire from the above equation that

$$\sigma(s_k)\left(\frac{q_k^{(j+1)} - q_k^{(j)}}{\tau} + \mathcal{O}(\tau)\right) - \frac{1}{a^2}\left(\frac{q_{k+1}^{(j)} - 2q_k^{(j)} + q_{k-1}^{(j)}}{h^2} + \mathcal{O}(h^2)\right) - \phi(q_k^{(j)}) = 0,$$

$$k = 1, 2, \ldots, N; \ j = 0, 1, \ldots, J.$$

Recall the Courant constraint $\tau/h^2 = \mathcal{O}(1)$ for thermodynamic finite difference approximations [18]. Then the above equalities imply that the pointwise defect

$$d_k^{(j)} = \sigma(s_k)\frac{q_k^{(j+1)} - q_k^{(j)}}{\tau} - \frac{1}{a^2}\frac{q_{k+1}^{(j)} - 2q_k^{(j)} + q_{k-1}^{(j)}}{h^2} - \phi(q_k^{(j)}) = \mathcal{O}(h^2), \quad (3.2)$$

$$k = 1, 2, \ldots, N; \ j = 0, 1, \ldots, J.$$

Therefore $\left\|d^{(j)}\right\|_\infty = \mathcal{O}(h^2)$ as $h \to 0^+$, $j \in \{0, 1, \ldots, J\}$. Hence the order of convergence of the data stream method (2.1), (2.2) is quadratic in the maximum norm. ∎

Remark 3.1. The order of convergence of the semi-adaptive simulation method (2.1), (2.2) is $r = 1.5$ in the Euclidean norm.

This can be seen readily from (3.2). We observe that $d_k^{(j)} \approx c_{k,j}h^2$, where $c_{k,j} > 0$ is a constant, $k = 1, 2, \ldots, N$; $j = 0, 1, \ldots, J$ Now,

$$\left\|d^{(j)}\right\|_2 \approx c_j\sqrt{N}h^4 = c_j\sqrt{N}h^2 \leq \frac{c_j}{\sqrt{h}}h^2 = c_jh^{3/2}, \ j = 0, 1, \ldots, J,$$

where c_j, $j = 0, 1, \ldots, J$, are positive constants. This completes our derivation.

Although a replacement definition based on defects cannot warrant (3.1), it offers straightforward estimate and is extremely convenient to use. In fact, the actual order of convergence defined in (3.1) is often lower than that calculated via the replacement definition [20,21]. Since a quadratic or higher convergence is often favorable to applications, such as those in cell migration model simulations [22], it is extremely meaningful to investigate and ensure the quality of convergence of (2.1), (2.2).

Fortunately, the task is possible through a generalized Milne device. This is stated through the following remark.

Remark 3.2. The order of convergence of the semi-adaptive simulation method (2.1), (2.2) is quadratic based on computational verifications via a generalized Milne device.

To demonstrate the result, we let q_h^τ denote the quenching solution with temporal steps $\tau \in \{\tau_1, \tau_2, \ldots, \tau_J\}$ h being the uniform spatial step. Further, let $q_{h/2}^\tau$ be the numerical solution with halved spatial step size $h/2$. Likewise, we define $q_{h/4}^\tau$ by a similar argument. It follows that a generalized Milne formula can be built for estimating the point-wise order of convergence via

$$
r_k^{(j)} = \frac{1}{\ln(2)} \ln \left| \frac{\left(q_k^{(2j)}\right)_{h/2}^\tau - \left(q_k^{(j)}\right)_h^\tau}{\left(q_k^{(4j)}\right)_{h/4}^\tau - \left(q_k^{(2j)}\right)_{h/2}^\tau} \right|, \quad k = 1, 2, \ldots, N; \ j = 1, 2, \ldots, J,
$$

(3.3)

given that the denominator is nontrivial [20]. Stretch the data from computational space $[0, 1]$ to original physical space $[0, 5]$ [19]. A surface of function $r_k^{(j)}$ is shown in Fig. 1. The surface can be viewed as a computational order of convergence spreading the entire space-time domain considered. Although the evaluation takes place on three "consecutive" meshes in the space, it can be conveniently extended for multidimensional cases. Detailed values of $r_k^{(j)}$ immediately before the quenching time are given in Table 1.

There is little surprise that such a surface of r obtained is not linear due to the strong quenching-blow up singularity of the quenching problem. We may notice the dramatic decay of the point-wise order of convergence from quadratic to one half in the quenching neighborhood near T_a. The phenomenon is consistent with existing discussions [15,19,20]. We note that if the variable Courant numbers $\gamma_j = \tau_j/h^2 \le a^2 \phi_{\min}$, $j = 1, 2, \ldots, J$, then a quadratic convergence in space implies a linear convergence in time, that is,

$$
\left\| q^{(j)} - q(t_j) \right\|_2 = \mathcal{O}(\tilde{\tau}), \quad \tilde{\tau} \to 0^+, \ j \in \{0, 1, \ldots, J\},
$$

(3.4)

where $\tilde{\tau} = \max_{0 \le j \le J} \tau_j$.

Formula (3.3) can be also extended via any standard p-norm, that is,

$$
r_\ell^{(j)} = \frac{1}{\ln(2)} \ln \frac{\left\| (q^{(2j)})_{h/2}^\tau - (q^{(j)})_h^\tau \right\|_p}{\left\| (q^{(4j)})_{h/4}^\tau - (q^{(2j)})_{h/2}^\tau \right\|_p}, \quad 1 \le p \le \infty, \ j = 1, 2, \ldots, J. \quad (3.5)
$$

On the other hand, if us consider a subset of the numerical solution sequence

$$
Q_{m,n} = \left\{ q^{(m)}, q^{(m+1)}, \ldots, q^{(j)}, \ldots, q^{(n)} \right\} \subseteq \left\{ q^{(0)}, q^{(1)}, \ldots, q^{(j)}, \ldots, q^{(J)} \right\},
$$

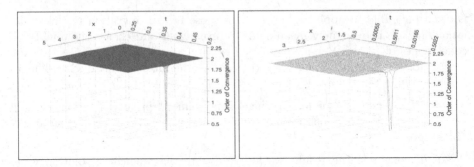

Fig. 1. [LEFT] Surface plot of the point-wise order of convergence of the numerical solution v by (2.1), (2.2). Formula (3.3) is employed.; [RIGHT] A locally enlarged surface plot of the left image near the quenching singularity. It can be observed that the the point-wise order is approximately quadratic except in a small area around the quenching data location which is at (s^*, T_a). Though the order decreases dramatically in such a small area, it still stays above 0.5 which indicates a satisfactory data reliability even in the tumor cell population blow-up area.

Table 1. The order of convergence of the solution $q(s, t)$ calculated via (3.3) at ten different s_j locations right before data quenching. Note the order is approximately 2 everywhere except around $s = 2.5$ which decreases to 0.5 due to the quenching singularity. These results are consistent with the existing theory [10–13].

s_j	order of convergence
2.02970297	1.99667070
2.12871287	1.99507152
2.22772277	1.99132454
2.32673267	1.97782804
2.42574257	1.80611369
2.52475247	0.50791656
2.62376237	1.95012804
2.72277227	1.98707873
2.82178217	1.99364935
2.92079207	1.99601168
3.01980198	1.99715469

where $0 \leq m < n \leq J$, then the following $(n - m + 1)$-dimensional vectors

$$w_k = \left(q_k^{(m)}, q_k^{(m+1)}, \ldots, q_k^{(j)}, \ldots, q_k^{(n)} \right)^{\mathsf{T}}, \quad k = 1, 2, \ldots, N,$$

can be defined. Therefore (3.5) can be simplified to

$$r_{\ell,k} = \frac{1}{\ln(2)} \ln \frac{\left\| w^{\tau}_{k,h/2} - w^{\tau}_{k,h} \right\|_{\ell}}{\left\| w^{\tau}_{k,h/4} - w^{\tau}_{k,h/2} \right\|_{\ell}}, \quad 1 \le \ell < \infty, \; k = 1, 2, \ldots, N. \qquad (3.6)$$

The above new formula is extremely convenient to use in cell simulation experiments. It also provides an effective order of convergence estimate spanning from t_m to t_n at each spacial location s_k, $k = 1, 2, \ldots, N$. It reflects the quality of performance of the algorithm dynamically in different stage of stream simulations.

Again, for the simplicity of illustration, let us consider the case with an uniform Courant number $\gamma = \tau/h^2 = 1.0201$ with $h = 1/101$. In Fig. 2, we present order estimates based on temporal intervals $[t_0, t_{1674}]$, $[t_{1675}, t_{3348}]$ and $[t_{3349}, t_{5023}]$ in Fig. 2. The spectral norm ($\ell = 2$) is used. It is found that the order is persistently stay at two, while decays repeatedly around the quenching-blow up location $s^* = 2.5$ in the last stage. We also notice that the order is slightly higher in the first stage probably due to the excellent stability of the argorithm.

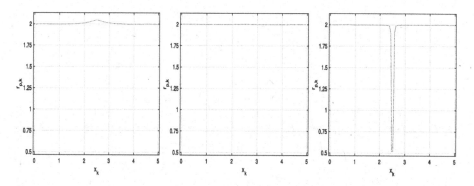

Fig. 2. Simulations of the order of convergence estimates based on formula (3.6). The three temporal stages used are $[t_0, t_{1674}]$, $[t_{1675}, t_{3348}]$ and $[t_{3349}, t_{5023}]$, respectively.

Finally, we may show a simulated quenching solution q and its rate-of-change function q_t. To do so, we may linearize (2.1), (2.2) to the following.

$$q^{(j+1)} = \left(I - \frac{\tau_j}{2} A\right)^{-1} \left(I + \frac{\tau_j}{2} A\right) \left[q^{(j)} + \frac{\tau_j}{2} \psi\left(q^{(j)}\right)\right] + \frac{\tau_j}{2} \psi\left(\omega^{(j+1)}\right), \; j = 0, 1, \ldots, J, \quad (3.7)$$

$$q^{(0)} = q_0, \qquad (3.8)$$

where

$$\omega^{(j+1)} \approx q^{(j+1)} = q^{(j)} + \tau_j \left[A q^{(j)} + \psi(q^{(j)})\right].$$

Consider a typical nonstochastic reaction function of the type (1.4) with $p = 1$. Let us keep $\tau^{(\ell)}$ to be uniformly for simplicity in simulations. A fixed physical space of $a = 5$ is selected.

Figure 3 shows the simulated solution and its corresponding temporal derivative function for the final 223 temporal levels before the quenching at $T_5 \approx 0.50111987$ ($J = 723$ temporal steps are used in the full simulation). Linearized scheme (3.7), (3.8) is utilized. A single point quench is observed at $s = 2.5$ as predicted [5, 10, 24]. The numerical solution is clearly nonnegative, and monotonically increasing as time t increases at any $s \in (0, a)$. It can also be observed that while the solution q remains bounded throughout the computation, its rate of change, that is, the temporal derivative function q_t seems to shoot to the infinity at $s = 2.5$ as data quenching is approached [9, 15, 22].

In the thermal physics, the phenomenon indicates that a combustion is ignited when the rate of change of the fuel temperature in a combustion chamber tends to be unbounded. In Table 2, we list maximal values of q and q_t in ten representative time levels, including six levels immediately before the quenching-blow up. We note that τ_j becomes variable after $j = 500$ due to the kick-in of the adaptation. The patterns of the data agree very well with those given in [5, 10].

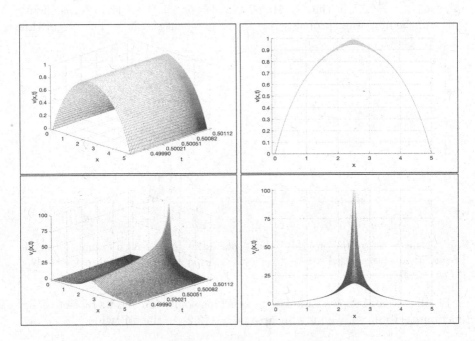

Fig. 3. Surface plots of the numerical solution q of (3.7)–(3.8) (first row), the corresponding rate derivative q_t (second row) for the last 223 temporal steps immediately before quenching. It can be observed that $T_5 \approx 0.50111987$, $\max\limits_{0 \le s \le 5, 0 \le t \le T_a} q(s, t) = q(2.5, T_a) \approx 0.99008661$ and $\max\limits_{0 \le s \le 5, 0 \le t \le T_a} q_t(s, t) = q_t(2.5, T_a) \approx 99.77142399$.

Our simulation strategy is the following. (i) A uniform time step τ is used until the solution almost quenches, for example, as the value $\max\limits_{0 \leq s \leq a} q(x,t)$ reaches 0.950. (ii) Then the sequence of adaptive time steps, $\{\tau_j\}$, begins through

$$\tau_j = \max\left\{\min_j\left\{\tau_{j-1}, c_0 \min_k\left\{\left(1 - q_k^j\right)^2\right\}\right\}, m_0\right\},$$

where $c_0 > 0$ is a suitable speed controller, and m_0 is a minimum step size that may keep the ratio of τ_j/τ_{j-1} being bounded and smooth [13,14]. A quadratic function is being used to reflect the nonlinearity and determine the next step size which allows the actual quenching singularity to drive the process. The above monitoring function developed is different from classical arc-length formulas and is highly satisfactory.

We plot the variable temporal steps generated as well as the performance ratio τ_j/τ_{j-1} in final 223 advancements in Fig. 4. It can be observed that while both q_j and $(q_t)_j$ increases monotonically, τ_k decreases monotonically due to our effective grid adaptation mechanism. In fact, the decay of τ_j is at a logarithmic rate.

Table 2. Maximal values of the solution $q(s,t)$ and its temporal derivative function $q_t(s,t)$ at ten different time levels before quenching-blow up. Note that the temporal adaptation starts at $j = 500$. Both values increase monotonically, with the latter increases exponentially immediately before the quenching-blow up. The results agree with known results given in [3,5,21,23].

j	t_j	$\max\limits_{0 \leq s \leq a} q(s,t_j)$	$\max\limits_{0 \leq s \leq a} q_t(s,t_j)$
10	0.01	0.00904086	1.00963768
100	0.1	0.10445547	1.11733601
500	0.49724053	0.91046311	11.22462715
700	0.50109449	0.98766458	81.72099321
715	0.50111365	0.98937234	94.91275358
716	0.50111474	0.98947743	95.86501759
717	0.50111580	0.98958149	96.82691747
718	0.50111685	0.98968452	97.79855187
719	0.50111787	0.98978654	98.78002048
720	0.50111887	0.98988755	99.77142399

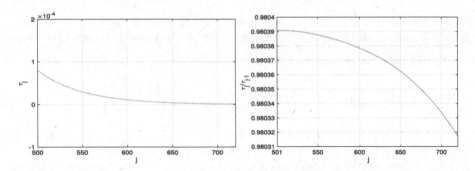

Fig. 4. [LEFT] Adaptive temporal step sizes used in the last 223 excursions of simulations. [RIGHT] Profile of the ratio τ_j/τ_{j-1}, in the final 223 temporal steps. It can be noticed that both τ_j and the ratio decrease monotonically while the $\|q_j\|_\infty$ increases monotonically.

4   Conclusions and Expectations

To conclude, in this paper, we have extended the theory and practice of semi-adaptive finite difference methods for degenerate quenching data stream simulations via nonlinear reaction-diffusion partial differential equations. The particular modeling equation studied is emerged from multiple bio-medical and physical applications, in particular in rapid cell bursts and solid fuel combustion [1, 2, 4, 7, 9, 13, 21, 22]. Systematic investigations and improved analysis of key characteristic issues, including the convergence, positivity and monotonicity of the simulated data streams.

Computational experiments are carried out to illustrate the theoretical results though demonstrations of the solution accuracy and order of convergences. It is found that the implicit simulation method implemented is quadratically convergent within the time-space physical regions considered.

The new preservative simulation method utilizes a uniform mesh in space and a temporal adaption in time. In our forthcoming work, we shall implement fully adaptive methods where mesh adaptations will be considered in both space and time. These may help further improve the accuracy and efficiency of the laboratory data. Geometrically non-symmetric degeneracy functions [10, 21] and initial values will be introduced and tested for potential biomedical and industrial applications.

References

1. Raghuraman, S., et al.: Pressure drives rapid burst-like coordinated cellular motion from 3D cancer aggregates. Adv. Sci. **9**, 2104808 (2022). https://doi.org/10.1002/advs.202104808
2. Butcher, D., Alliston, T., Weaver, V.: A tense situation: forcing tumour progression. Nat. Rev. Cancer **9**, 108–122 (2009). https://doi.org/10.1038/nrc2544
3. Kawarada, H.: On solutions of initial-boundary value problems for $u_t = u_{xx} + 1/(1-u)$. Publ. Res. Inst. Math. Sci. **10**, 729–736 (1975)

4. Poinset, T., Veynante, D.: Theoretical and Numerical Combustion, 2nd edn. Edwards Publisher, Philadelphia (2005)
5. Sheng, Q., Khaliq, A.: A revisit of the semi-adaptive method for singular degenerate reaction-diffusion equations, East Asia. J. Appl. Math. **2**, 185–203 (2012)
6. Beauregard, M., Sheng, Q.: An adaptive splitting approach for the quenching solution of reaction-diffusion equations over nonuniform grids. J. Comp. Appl. Math. **241**, 30–44 (2013)
7. Bebernes, J., Eberly, D.: Mathematical Problems from Combustion Theory. Springer, New York (1989). https://doi.org/10.1007/978-1-4612-4546-9
8. Bon, T.K., Kouakou, T.K.: Continuity of the quenching time in a semilinear heat equation with a potential. Revista Colombiana de Matemáticas **43**, 55–70 (2009)
9. Zirwes, T., et al.: Numerical study of quenching distances for side-wall quenching using detailed diffusion and chemistry. Flow Turb. Combustion **106**, 649–679 (2021)
10. Acker, A.F., Kawohl, B.: Remarks on quenching. Nonlinear Anal. Theory Meth. Appl. **13**, 53–61 (1989)
11. Hale, J.K.: Asymptotic Behavior of Dissipative Systems. American Math Soc, Philadelphia (1988)
12. Levine, H.A.: Quenching, nonquenching, and beyond quenching for solution of some parabolic equations. Annali di Matematica **155**, 243–260 (1989)
13. Padgett, J.L., Sheng, Q.: Nonuniform Crank-Nicolson scheme for solving the stochastic Kawarada equation via arbitrary grids. Numer. Meth. PDEs **33**, 1305–1328 (2017)
14. Huang, W., Russell, R.D.: Adaptive Moving Mesh Methods. Springer, Zürich (2011)
15. Nouaili, N.: A Liouville theorem for a heat equation and applications for quenching. Nonlinearity **24**, 797–832 (2011)
16. Cheng, H., Lin, P., Sheng, Q., Tan, R.: Solving degenerate reaction-diffusion equations via variable step Peaceman-Rachford splitting. SIAM J. Sci. Comput. **25**(4), 1273–1292 (2003)
17. Sheng, Q., Torres, E.S.: A nonconventional stability approach for a nonlinear Crank-Nicolson method solving degenerate Kawarada problems (2023, submitted and under reviews)
18. Iserles, A.: A First Course in the Numerical Analysis of Differential Equations, 2nd edn. Cambridge University Press, Cambridge and London (2009)
19. Padgett, J.L., Sheng, Q.: On the positivity, monotonicity, and stability of a semi-adaptive LOD method for solving three-dimensional degenerate Kawarada equations. J. Math. Anal. Appls. **439**, 465–480 (2016)
20. Kabre, J., Sheng, Q.: A preservative splitting approximation of the solution of a variable coefficient quenching problem, Computers. Math. Appl. **100**, 62–73 (2021)
21. Sheng, Q., Khaliq, A.: Integral Methods in Science and Engineering (Research Notes in Math), Chapman and Hall/CRC, London and New York (2001). Ch. 9. A monotonically convergent adaptive method for nonlinear combustion problems
22. D. Krndija, et al.: Active cell migration is critical for steady-state epithelial turnover in the gut, Science 365 (2019) 705–710. https://doi.org/10.11268/science.aau3429
23. Zhou, J.: Quenching for a parabolic equation with variable coefficient modeling MEMS technology. App. Math. Comput. **314**, 7–11 (2017)
24. Liang, K., Lin, P., Tan, R.: Numerical solution of quenching problems using mesh-dependent variable temporal steps. App. Numer. Math. **57**, 791–800 (2007)

A Study of a Metapopulation Model Using the Stochastic Reaction Diffusion Master Equation

Md Mustafijur Rahman◉ and Roger B. Sidje(✉)◉

Department of Mathematics, The University of Alabama, Tuscaloosa, AL 35487, USA
mrahman24@crimson.ua.edu, roger.b.sidje@ua.edu

Abstract. Many biological systems involve both the noise in chemical reaction process and diffusion in space between the molecules involved in the reactions. Since the copy numbers of participating species often are small, their stochastic behavior comes into play. The reaction-diffusion master equation (RDME) is a stochastic modeling of the reaction-diffusion process. In RDME, the space is partitioned into compartments and a well mixing of species is considered inside each compartment. Contrary to the chemical master equation (CME) that does not analyze the diffusion of the molecules between the components, the RDME is more complex and has a substantially larger state space than the CME. In this study we tackle a metapopulation model using the RDME formulation. Numerical methods are used to analyze and predict the behavior of the model.

Keywords: metapopulation · reaction diffusion master equation · chemical master equation · stochastic simulation algorithm

1   Introduction

Biological systems at the cellular level involve random interactions among species that have small copy numbers. This fact has been well-established in the area of molecular cell biology through both experimental investigations and mathematical modeling. The spatial environment within biological cells is highly intricate and consists of geometrically complex structures. The chemical master equation (CME) [15] is a mathematical framework used to describe the stochastic behavior of chemical reactions in a well-mixed system. It provides a way to model the time evolution of the probability distribution of the number of molecules of each species in a system, taking into account both the discrete nature of molecules and the stochasticity of their interactions.

In recent years, there has been considerable attention given to the development of effective computational techniques for performing discrete stochastic simulation of well-mixed biochemical systems [16]. The stochastic simulation algorithm (SSA) [13,14] is a computational method employed to obtain sample realizations guided by the chemical kinetics that underpin the CME. The

H. Han and E. Baker (Eds.): SDSC 2023, CCIS 2113, pp. 242–253, 2024.
https://doi.org/10.1007/978-3-031-61816-1_17

SSA is a powerful tool for simulating complex biochemical networks, as it can take into account the stochasticity of the system and can generate statistically meaningful results. We refer to [18] for a demonstration of an implementation to the Michaelis-Menten system. Despite being useful, the SSA algorithm is time consuming since it is based on realizations of the states over time. Acceleration techniques such as tau-leaping have been introduced to make this algorithm faster [20]. The finite state projection algorithm (FSP) [23] is another way of tackling the CME where it involves solving a set of differential equations that describe the change in the probability distribution over time. The equations are based on the transition rates between states in the system and the current probability distribution. The algorithm can be used to compute the probability of reaching a specific state or a set of states within a given time interval.

Unlike traditional chemical reaction models that assume that reactions occur instantaneously and uniformly throughout the system, cellular environments are not uniform in space, and spatial heterogeneity is a common feature. In numerous biochemical systems, the spatial heterogeneity of species cannot be disregarded, rendering the systems not well-stirred. This may be due to the slow transport of molecules through the solvent compared to typical reaction times or the strong localization of some reactions. Experimental data from various sources [7,8,22], demonstrate the significance of considering both the stochastic characteristics and spatial distribution of a system to explain its behavior. Several mathematical methods, which are different but interrelated [2,9,31], have recently been employed to describe stochastic reaction-diffusion systems in biological cells [25,30]. Here, the assumption is that molecules in stochastic reaction-diffusion systems behave as points that undergo continuous Brownian motion in space. Whenever two molecules come within a defined reaction radius, bimolecular chemical reactions occur instantly. This modeling approach is referred to as spatially continuous diffusion-limited reaction (SCDLR), which was originally proposed by Smoluchowski [27]. In contrast to these methods, when we treat the diffusion at a molecular level as a special set of reactions in the CME, we arrive at the reaction-diffusion master equation (RDME) [11].

RDME is a mathematical framework used to describe the dynamics of stochastic chemical reaction systems that also involve diffusion processes. The equation describes the time-dependent probability distribution function (PDF) of the system state, similar to the CME, but with a significantly higher dimensionality. In RDME, the space is divided into compartments. Each compartment is considered to be well-mixed and reactions within a particular compartment are considered to be consistent with the homogeneous case. Also, molecules can get diffusive transfers between adjacent compartments. The state of the system at any given time is defined by the number of molecules of each species in each compartment. Mathematically, the RDME is the forward Kolmogorov equation for a continuous-time jump Markov process. To tackle the RDME one feasible way is to generate samples from the RDME and collect statistics in a Monte Carlo fashion. The works in [4,28] illustrate some examples of the SSA applied to reaction-diffusion systems in one space dimension. The next reaction method (NRM) [12] is a version of the SSA that is known for its efficiency. A software called MesoRD [17] offers a specialized implementation of NRM, which is designed for diffusive systems and known as the next subvolume method (NSM) [9]. The aim of the binomial tau-leap spatial stochastic simulation algorithm [21] is to enhance its performance by merging the concepts of consolidating diffusive transfers and utilizing the priority queue structure from the NSM. In certain situations, it may be feasible to consider diffusion in a deterministic manner, thereby minimizing the need to monitor fast diffusive transfers to a great extent. The reactions are handled by SSA in those situations. This approach is exemplified by the hybrid multiscale kinetic Mónte Carlo method [29] and the Gillespie multiparticle method [26]. The corresponding macroscopic equation for the concentrations of the species is the reaction-diffusion equation (RDE) which is a partial differential equation (PDE). This equation is employed in diverse applications and is valid for large numbers of molecules when stochastic effects can be neglected.

Here, we formulate a metapopulation model using the RDME. Our study employs the FSP technique to the metapopulation model. Although the RDME model is similar with the CME, it is computationally more expensive since the state space becomes substantially larger. We have applied the SSA algorithm to analyze the marginal distribution of the species involved. Additionally, we envision a potential remedy to expedite the algorithm by using tensors.

2   Background

2.1   CME

Consider a system involving N molecular species $\{S_1, \ldots, S_N\}$, represented by the state vector $\mathbf{X}(t) = [X_1(t), \ldots, X_N(t)]^T$, where each $X_i(t)$ is a non-negative integer denoting the number of molecules of species S_i at time t. The state space of the system is the set of distinct vectors that correspond to the possible configurations the system can have. As many types of reactions occur, the system evolves from state to state through an underlying Markov chain. Suppose there are M reaction channels, denoted by $\{R_1, \ldots, R_M\}$, and assume that the system is well mixed and in thermal equilibrium. The dynamics of the reaction channel R_j is characterized by two factors: the propensity function a_j, which describes the likelihood of the reaction occurring given the current state of the system, and the state change (stoichiometric) vector $\boldsymbol{\nu}_j = [\nu_{1j}, \ldots, \nu_{Nj}]^T$, which specifies the change in the number of molecules of each species that results from one instance of the reaction; $a_j(\mathbf{x})dt$ gives the probability that, given $\mathbf{X}(t) = \mathbf{x}$, one R_j reaction will occur in the next infinitesimal time interval $[t, t + dt]$, and ν_{ij} gives the change in X_i induced by one R_j reaction. That means the components of the state change vector are integers displaying the increase or decrease of the number of copies of each species after the associated reaction occurs.

This system can be modeled as a Markov process, with its behavior determined by the Chemical Master Equation (CME) [15]

$$\frac{dP(\mathbf{x},t)}{dt} = \sum_{j=1}^{M} [a_j(\mathbf{x}-\boldsymbol{\nu}_j)P(\mathbf{x}-\boldsymbol{\nu}_j,t) - a_j(\mathbf{x})P(\mathbf{x},t)], \tag{1}$$

where the function $P(\mathbf{x},t)$ denotes the probability that $\mathbf{X}(t)$ will be $\mathbf{x}$. Equation (1) may be written in an equivalent matrix-vector form by enumerating all the states. If there are n possible states, $\mathbf{x}_1,\ldots,\mathbf{x}_n$, the CME takes the form of a system of linear ordinary differential equations (ODEs),

$$\dot{\mathbf{P}}(t) = \boldsymbol{\mathcal{M}} \cdot \mathbf{P}(t), \tag{2}$$

where $\boldsymbol{\mathcal{M}}$ is the transition matrix that describes the chemical reactions and the probability vector $\mathbf{P} = [p_1,\ldots,p_n]^T$ is such that each component $p_i = P(\mathbf{x}_i,t) = Prob(\mathbf{x}(t) = \mathbf{x}_i)$, the probability of being at state $\mathbf{x}_i$ at time t, for $i = 1,\ldots,n$. Computational challenges immediately arise for systems with many reactions or species. This is because the CME requires the numerical solution of a set of differential equations, which can be computationally intensive. As a result, we often use approximations or simplified versions of the CME to make calculations more manageable.

2.2   SSA

The "curse of dimensionality" obstructs finding an analytical solution to the CME for larger systems. This arises because the number of possible states increases exponentially as the number of molecular species and reactions increases. To overcome this limitation, the SSA is widely used as a numerical method to simulate the time evolution of chemical reaction networks. This method is based on the principles of the CME and generates a stochastic trajectory of the network by simulating individual reactions one by one. SSA selects the next reaction and time interval by calculating the propensity functions of all possible reactions based on the current state of the system, generating a random number to determine the time until the next reaction occurs, and selecting the next reaction based on probabilities of each reaction occurring. The algorithm updates the system state according to the reaction stoichiometry and repeats the process.

The SSA produces a pair of random numbers, namely r_1 and r_2, during every individual step. Note that r_1 and r_2 both belong to $U(0,1)$, which is the set of uniformly distributed random numbers in the interval $(0,1)$. The time for the next reaction to occur is given by $t + \tau$, where τ is given by

$$\tau = \frac{1}{a_0} \ln\left(\frac{1}{r_1}\right). \tag{3}$$

It is now important to know which reaction will occur next. The index λ of the next reaction is given by the smallest integer which satisfies the following inequality,

$$\sum_{j=1}^{\lambda} a_j > r_2 a_0, \tag{4}$$

where,

$$a_0(\mathbf{x}) = \sum_{j=1}^{M} a_j(\mathbf{x}). \tag{5}$$

The states of the system are updated by $\mathbf{X}(t+\tau) = \mathbf{X}(t) + \boldsymbol{\nu}_\lambda$. Subsequently, the simulation proceeds to the time of the next reaction. By generating a large number of stochastic trajectories, the SSA method can provide an accurate approximation of the behavior of the reaction network over time, even in cases where an analytical solution is not feasible due to the curse of dimensionality.

2.3 FSP

Instead of simulating an ensemble of trajectories using SSA, the FSP method [23] directly computes an analytical approximation to the solution of the CME. This is achieved by creating a computationally manageable projection of the full state space and calculating the time evolution of the probability mass function within this projection space. The FSP technique uses a truncated state space to solve the CME and estimate the probability vector (PV) of the populations in a chemical reaction system. For a truncated transition matrix $\mathcal{M}_T$ and initial truncated PV denoted by $\mathbf{P}_T(t=0)$, the FSP finds the PV at any time $\mathbf{P}_T(t)$, by the following,

$$\dot{\mathbf{P}}_T(t) = \mathcal{M}_T \cdot \mathbf{P}_T(t). \tag{6}$$

Equation (6) is a system of linear ODEs and the solution is given by

$$\mathbf{P}_T(t) = \exp(\mathcal{M}_T t) \mathbf{P}_T(0). \tag{7}$$

2.4 RDME

Assume the domain Ω is partitioned into compartments (voxels). We label the compartments with $V_k, k = 1, \ldots, K$. Molecules within each compartment can react with one another within that compartment, and they can also diffuse across the boundaries and move to neighboring compartments. Both the reaction and diffusion processes are considered as random processes. Let, $X_{i,k}(t)$ be the number of molecules of species S_i in compartment V_k at time t. Then each species in the domain is given by the subvector $\mathbf{X}_i(t) = [X_{i,1}(t), \ldots, X_{i,K}(t)], i = 1, \ldots, N$. That means $\mathbf{X}_1$ is the subvector which represents the species 1 in all compartments, $\mathbf{X}_2$ represents species 2, and so on. Thus the state vector of the system is $\mathbf{X} = [\mathbf{X}_1, \ldots, \mathbf{X}_N]$. The diffusion propensity function $d_{i,j,k}$ and the state change vector $\boldsymbol{\mu}_{k,j}$ characterize the dynamics of the diffusion of species S_i from compartment V_k to V_j. The vector $\boldsymbol{\mu}_{k,j}$ has a length of K with -1 in the kth position, 1 in the jth position, and 0 elsewhere. Given, $\mathbf{X}(t) = \mathbf{x}$ the diffusion master equation (DME) can be written by

$$\frac{dP(\mathbf{x},t)}{dt} = \sum_{i=1}^{N} \sum_{k=1}^{K} \sum_{j=1}^{K} [d_{i,j,k}(\mathbf{x}_i - \boldsymbol{\mu}_{k,j}) P(\mathbf{x}_1, \ldots, \mathbf{x}_i - \boldsymbol{\mu}_{k,j}, \ldots, \mathbf{x}_N, t)$$
$$- d_{i,j,k}(\mathbf{x}_i) P(\mathbf{x}, t)]. \tag{8}$$

The diffusion propensity function $d_{i,j,k}(\mathbf{x}_i) = D/l^2$, for $k = j \pm 1$, where D is the diffusion rate and l is the characteristic length of the compartment. If $\mathcal{D}$ is the transition matrix describing the diffusion of molecules, the equivalent matrix-vector form can be written by

$$\dot{\mathbf{P}}(t) = \mathcal{D} \cdot \mathbf{P}(t). \tag{9}$$

Combining Eqs. (2) and (9) we get the matrix-vector form of the RDME,

$$\dot{\mathbf{P}}(t) = \mathcal{M} \cdot \mathbf{P}(t) + \mathcal{D} \cdot \mathbf{P}(t). \tag{10}$$

Equation (10) is a system of linear constant coefficient ODEs and gives us more possible states than the CME, and so its corresponding $\mathcal{M}$ is substantially extended to represent species in compartments. Overall, the RDME captures both the reaction and diffusion hence it has a much higher dimensionality.

3   Results

3.1   Metapopulation Model

Metapopulation models, which are used to study populations that are divided into subpopulations connected by migration, often employ reaction-diffusion models to describe spatial dynamics [5]. In a reaction-diffusion metapopulation model, there is a diffusion term and a reaction term. The diffusion term captures the movement of individual between neighboring subpopulations, while the reaction term describes birth and death rates within each subpopulation. The reaction-diffusion model has many applications, such as predicting how changes in habitat quality or connectivity may affect a species' distribution and abundance, studying disease spread, and assessing the persistence of rare or endangered species [3,10]. By taking into account both diffusion and reaction terms, the model can better capture complex interactions between local population dynamics and spatial dispersal.

In a metapopulation model using the RDME, each subpopulation's population is modeled as a stochastic process with birth, death, and migration events occurring randomly. The model includes diffusion of individuals between subpopulations and reactions within each subpopulation, such as birth and death rates. The RDME takes into account the stochasticity of these events, allowing for more accurate predictions of population dynamics compared to deterministic models.

For the purpose of experimentally demonstrating a proof-of-concept in this study, we consider a basic reaction scheme conserving the number of particles

$$\beta \xrightarrow{c_1} \alpha, \tag{11}$$

$$\alpha + \beta \xrightarrow{c_2} 2\beta. \tag{12}$$

This system has been widely used both in physics and mathematical epidemiology. For its interpretation with the lenses of our metapopulation model, it would be that α particles represent normal particles (healthy individuals) and β particles represent active particles (infected individuals) intermingling from area to area. This scheme is also known as the SIS model [1]. We consider two neighboring areas (i.e., compartments) where both the healthy and infected individuals can move back and forth. We label the areas as Area 1 and Area 2. Hence the scheme can be represented by a reaction-diffusion process where the reactions (11) and (12) can take places inside both areas separately and the individuals can move back and forth between Area 1 and Area 2. The process is a random process and can be modeled by the RDME consisting of two reactions with two compartments. The total number of particles in the process remains constant because the particles undergo a transformation from one form to another without being created or destroyed. To allow visualizing our prototype, we assume that initially, there are 2 healthy individuals and 1 infected individual in area 1. Similarly, in area 2 there are 1 healthy and 2 infected individuals. The initial state vector is $[2, 1, 1, 2]^T$, and we set the reaction parameters $c_1 = .30$ and $c_2 = 1.0$, and the diffusion parameters $D = 1.0$ and $l = 10$. The RDME gives us all the possible states we can get when those individuals react and diffuse randomly. Even such a simple initial state generates 84 possible states.

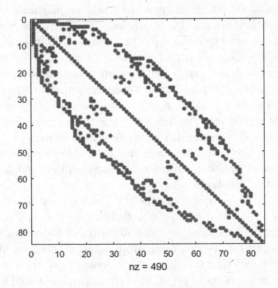

Fig. 1. Visualization of the sparsity pattern of the transition matrix in RDME.

Figure 1 shows the sparsity pattern of the transition matrix involved in the RDME formulation where the transition matrix covers both reaction and diffusion process. In the CME, the transition matrix would have only been responsible for the reactions. In the RDME the matrix we have is a 84×84 matrix, where the dots in the Fig. 1 represent the 490 non-zero elements of the transition matrix and the other parts without dots indicate zeros. As noted earlier, this example is a proof-of-concept with a resulting figure that a reader can easily visualize. More realistic models lead to substantially larger sizes. Indeed, for a problem with N species S_i that each can attain 0 to $n_i - 1$ copies, the CME matrix can already be of size $\prod_{i=1}^{N} n_i \times \prod_{i=1}^{N} n_i$, so that if $N = 10$ and $n_i \approx 10$ the size would be a huge $10^{10} \times 10^{10}$ for the CME and even more than that for the RDME.

We solve the RDME using the FSP Eq. (7) by truncating the state space to 70, to approximate the probability of the states. We are particularly interested to get the marginal probability of the individuals. By definition, the marginal probability of an event is the likelihood of an event happening regardless of the outcome of the other events. Assuming $t = 10$, we approximate the marginal probability distribution of the healthy and infected individuals in two areas. The results are shown in Fig. 2

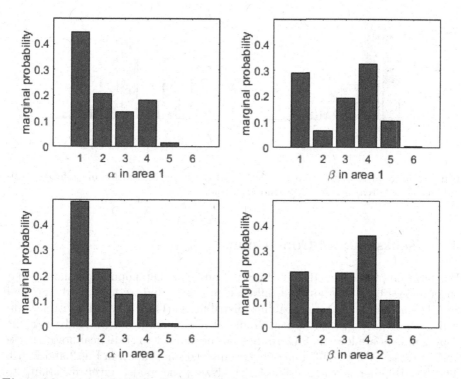

Fig. 2. Marginal probability distribution of the number of healthy and infected individuals in two areas using FSP at time $t = 10$.

Another algorithm of analyzing the model is using the SSA. Contrary to the FSP, the SSA generates realizations. The computational time of this algorithm depends on the number of realizations taken into account. Here we have taken 100 realizations to find the marginal probability distribution of the individuals, with the result shown in Fig. 3.

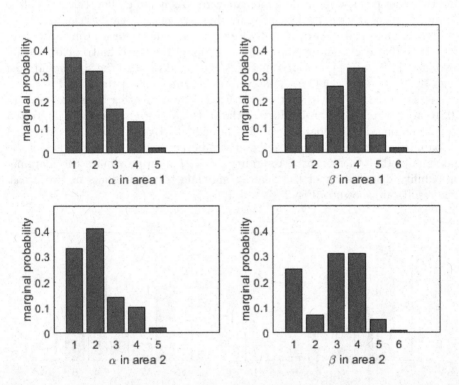

Fig. 3. Marginal probability distribution of the number of healthy and infected individuals in two areas using SSA at time $t = 10$.

4   Discussion and Conclusion

We have employed the RDME formulation in a metapopulation model that encompasses both reaction and diffusion processes and we have used FSP and SSA to find the marginal probability distribution. The FSP works with a truncated state space instead of considering all the possible states. And when the copy numbers are low the SSA-results can be subject to statistical noise. These are the reasons for the differences between the results shown in Fig. 2 and Fig. 3. The RDME approach is suitable for analyzing the model, given its ability to relate the model, even with the challenge of the "curse of dimensionality". The curse of dimensionality refers to the fact that as the dimensionality of the model increases, the number of possible states grows exponentially, making it increasingly difficult to calculate the probabilities of future states. In addition, implementing the FSP in the RDME or using the SSA algorithm to approximate the probability of future states can prove to be computationally expensive. In cases where the metapopulation model is more complex and involves an exceedingly vast state space, the computational time required to solve it using either of these methods becomes prohibitively high. As such, the main objective of our future work is to overcome these challenges and find a more efficient way to solve the RDME for advanced metapopulation models. To achieve this goal, we envision the use of tensor train decomposition techniques [24]. Their basic idea is to represent a high-dimensional tensor as a sequence of smaller tensors that are multiplied together. This sequence of smaller tensors is called the tensor train. Each tensor in the sequence has a fixed size and is connected to the neighboring tensors by a set of indices. Applications of tensors already exist in the context of the CME [6,19], and we similarly envision that they could provide an alternative approach to calculate the RDME solution that can significantly reduce computational time while still maintaining accuracy, thereby enabling us to meaningfully analyze and model metapopulation systems and gain insights into their dynamics and behavior. Overall, a successful implementation of the tensor train decomposition technique in the context of the RDME formulation for metapopulation modeling could pave the way for more efficient and accurate modeling of complex systems. This, in turn, could have significant applications in various fields such as ecology, biology, and epidemiology. By improving our understanding of these systems, we can better predict and mitigate the spread of diseases, preserve ecological systems, and ensure the long-term sustainability of biological populations.

References

1. Anderson, R.M., May, R.M., Anderson, B.: Infectious Diseases of Humans: Dynamics and Control. Oxford University Press, Oxford (1992)
2. Andrews, S.S., Bray, D.: Stochastic simulation of chemical reactions with spatial resolution and single molecule detail. Phys. Biol. **1**(3), 137 (2004)
3. Baronchelli, A., Catanzaro, M., Pastor-Satorras, R.: Bosonic reaction-diffusion processes on scale-free networks. Phys. Rev. E **78**(1), 016111 (2008)
4. Bernstein, D.: Simulating mesoscopic reaction-diffusion systems using the Gillespie algorithm. Phys. Rev. E **71**(4), 041103 (2005)
5. Colizza, V., Pastor-Satorras, R., Vespignani, A.: Reaction-diffusion processes and metapopulation models in heterogeneous networks. Nat. Phys. **3**(4), 276–282 (2007)
6. Dinh, T., Sidje, R.B.: An adaptive solution to the chemical master equation using quantized tensor trains with sliding windows. Phys. Biol. **17**(6), 065014 (2020)
7. Dobrzyński, M., Rodríguez, J.V., Kaandorp, J.A., Blom, J.G.: Computational methods for diffusion-influenced biochemical reactions. Bioinformatics **23**(15), 1969–1977 (2007)
8. Doubrovinski, K., Howard, M.: Stochastic model for soj relocation dynamics in bacillus subtilis. Proc. Nat. Acad. Sci. **102**(28), 9808–9813 (2005)

9. Elf, J., Ehrenberg, M.: Spontaneous separation of bi-stable biochemical systems into spatial domains of opposite phases. Syst. Biol. **1**(2), 230–236 (2004)

10. Ferreira, S.C., Martins, M.L.: Critical behavior of the contact process in a multi-scale network. Phys. Rev. E **76**(3), 036112 (2007)

11. Gardiner, C., McNeil, K., Walls, D., Matheson, I.: Correlations in stochastic theories of chemical reactions. J. Stat. Phys. **14**, 307–331 (1976)

12. Gibson, M.A., Bruck, J.: Efficient exact stochastic simulation of chemical systems with many species and many channels. J. Phys. Chem. A **104**(9), 1876–1889 (2000). https://doi.org/10.1021/jp993732q

13. Gillespie, D.T.: A general method for numerically simulating the stochastic time evolution of coupled chemical reactions. J. Comput. Phys. **22**(4), 403–434 (1976)

14. Gillespie, D.T.: Exact stochastic simulation of coupled chemical reactions. J. Phys. Chem. **81**(25), 2340–2361 (1977)

15. Gillespie, D.T.: A rigorous derivation of the chemical master equation. Phys. A **188**(1–3), 404–425 (1992)

16. Gillespie, D.T.: Stochastic simulation of chemical kinetics. Annu. Rev. Phys. Chem. **58**(1), 35–55 (2007). https://doi.org/10.1146/annurev.physchem.58.032806. 104637. pMID: 17037977

17. Hattne, J., Fange, D., Elf, J.: Stochastic reaction-diffusion simulation with MesoRD. Bioinformatics **21**(12), 2923–2924 (2005)

18. Higham, D.J.: Modeling and simulating chemical reactions. SIAM Rev. **50**(2), 347–368 (2008)

19. Kazeev, V., Khammash, M., Nip, M., Schwab, C.: Direct solution of the chemical master equation using quantized tensor trains. PLoS Comput. Biol. **10**(3), e1003359 (2014)

20. Marquez-Lago, T.T., Burrage, K.: Binomial tau-leap spatial stochastic simulation algorithm for applications in chemical kinetics. J. Chem. Phys. **127**(10), 09B603 (2007)

21. Marquez-Lago, T.T., Burrage, K.: Binomial tau-leap spatial stochastic simulation algorithm for applications in chemical kinetics. J. Chem. Phys. **127**(10), 104101 (2007). https://doi.org/10.1063/1.2771548

22. Metzler, R.: The future is noisy: the role of spatial fluctuations in genetic switching. Phys. Rev. Lett. **87**(6), 068103 (2001)

23. Munsky, B., Khammash, M.: The finite state projection algorithm for the solution of the chemical master equation. J. Chem. Phys. **124**(4), 044104 (2006)

24. Oseledets, I.V.: Tensor-train decomposition. SIAM J. Sci. Comput. **33**(5), 2295–2317 (2011)

25. Pavin, N., Paljetak, H.Č, Krstić, V.: Min-protein oscillations in escherichia coli with spontaneous formation of two-stranded filaments in a three-dimensional stochastic reaction-diffusion model. Phys. Rev. E **73**(2), 021904 (2006)

26. Rodriguez, J.V., Kaandorp, J.A., Dobrzyński, M., Blom, J.G.: Spatial stochastic modelling of the phosphoenolpyruvate-dependent phosphotransferase (PTS) pathway in escherichia coli. Bioinformatics **22**(15), 1895–1901 (2006)

27. Smoluchowski, M.: Mathematical theory of the kinetics of the coagulation of colloidal solutions. Z. Phys. Chem. **92**, 129–168 (1917)

28. Stundzia, A.B., Lumsden, C.J.: Stochastic simulation of coupled reaction-diffusion processes. J. Comput. Phys. **127**(1), 196–207 (1996)

29. Zheng, Z., Stephens, R.M., Braatz, R.D., Alkire, R.C., Petzold, L.R.: A hybrid multiscale kinetic Monte Carlo method for simulation of copper electrodeposition. J. Comput. Phys. **227**(10), 5184–5199 (2008)

30. van Zon, J.S., Morelli, M.J., Tănase-Nicola, S., ten Wolde, P.R.: Diffusion of transcription factors can drastically enhance the noise in gene expression. Biophys. J . **91**(12), 4350–4367 (2006)
31. van Zon, J.S., Ten Wolde, P.R.: Simulating biochemical networks at the particle level and in time and space: Green's function reaction dynamics. Phys. Rev. Lett. **94**(12), 128103 (2005)

Author Index

H. Han and E. Baker (Eds.): SDSC 2023, CCIS 2113, pp. 255–256, 2024.
https://doi.org/10.1007/978-3-031-61816-1

Printed in the United States
by Baker & Taylor Publisher Services